for
Michael Murray

RECENT EARTH MOVEMENTS

an introduction to neotectonics

C. VITA-FINZI

University College London

1986

ACADEMIC PRESS

Harcourt Brace Jovanovich, Publishers

London · Orlando · San Diego · New York
Austin · Montreal · Sydney · Tokyo · Toronto

ACADEMIC PRESS INC. (LONDON) LTD.
24-28 Oval Road, London NW1 7DX

United States Edition published by
ACADEMIC PRESS, INC.
Orlando, Florida 32887

British Library Cataloguing in Publication Data

Vita-Finzi, C.
Recent earth movements: an introduction to
neotectonics.
1. Geology, Structural
I. Title
551.1'36   QE601.2

ISBN 0-12-722370-3 (casebound)
ISBN 0-12-722371-1 (paperback)

Phototypeset by
Dobbie Typesetting Service, Plymouth, Devon

Printed in Great Britain
at the University Press, Cambridge

CONTENTS

INTRODUCTION

Evidence for Movement

Potential Value

potential
value

LIST OF ILLUSTRATIONS

PREFACE

In this book I illustrate the main sources of evidence for recent earth movements and try to show why the evidence is worth collecting. Two kinds of crustal movements are doubtless familiar to most readers: the ground breaks associated with large earthquakes, and the gradual uplift that has affected Scandinavia and Canada since the Pleistocene ice sheets retreated. These were once regarded as anomalous phenomena in an otherwise stable earth. Nowadays the anomaly is on the other foot: no part of the globe can safely be declared terra firma. As I argue in the introductory chapter, the reason lies in changed attitudes no less than in novel data. It is a sign of the times that the International Association for Earthquake Engineering now has an active British section.

Although I write mainly for undergraduate students of geology and geomorphology, I hope what follows will also interest historians, engineers and curious travellers. Recent deformation has some bearing (sometimes through not having occurred) on the interpretation of human affairs; it is proving an essential ingredient of attempts to trace and interpret the pattern of earthquake activity in space and time; and its study still stands to benefit from disinterested reporting.

Recent is used in the title as it was in *Recent Earth History* (Macmillan, 1973) in its informal sense to denote the immediate rather than the distant past, although I take refuge in the technical definition of the geological term Recent—viz. the last 10 000 years—when a time limit becomes desirable. The word 'neotectonics' gave me more trouble: it is pretentious and in some ways misleading, yet it serves to define the subject and has gained wide currency. In the end I used it for the subtitle.

The geological, archaeological and instrumental evidence for movement is considered in Chapters 2–5. I settled for this approach, rather than for a discussion of different kinds of geological structure, mainly because one can be certain that movement has taken place without necessarily understanding its precise nature let alone the reasons for it. Chapters 6–9 consider the potential value of the information for various fields of research. By the time I had written them I was quite persuaded by my case. At the very least the examples I cite should prompt the reader into looking at the landscape in a fresh way.

A full list of sources reveals what I have read and doubtless also the many interesting items I did not know about. Most of the books and journals quoted are relatively unobscure.

ACKNOWLEDGEMENTS

Of the colleagues who guided me in the field I am especially indebted to J. A. Jackson, G. C. P. King, D. J. Shearman and R. H. Sibson. Parts of the book were read in draft by J. E. Guest (who also supplied four of the photographs), R. Muir Wood and Dr King. R. S. White supplied Fig. 75. Mrs L. Copeland identified many artefacts and drew those on Fig. 29. Miss A. Swindells prepared the manuscript, Miss S. Gatsky drew the figures and C. Cromarty printed most of the photographs. I thank them all.

I am much indebted to the following, as well as to the authors in question, for permission to use copyright material in the preparation of figures drawn for this book: The American Association for the Advancement of Science (Figs 3, 39, 52, 54, 89, 93, 102 and 111); The American Geophysical Union (Figs 11, 88, 99 and 100); Cambridge University Press (Figs 8 and 41); W. H. Freeman & Co. (Figs 4 and 108); The Geological Society of America (Figs 18, 24, 26 and 101); Macmillan (Fig. 69); Macmillan Journals Ltd (Figs 12, 35, 58, 63, 86 and 110); Methuen & Co. (Fig. 2); John Wiley & Sons Ltd (Figs 6 and 85). American Association of Petroleum Geologists (Fig. 75); the Editor, *Boll. geofis. teor. appl.* (Fig. 19); the Director, Bureau of Mineral Resources, Canberra (Figs 65 and 109); Elsevier Science Publishers (Figs 5, 65 and 66); Gebrüder Borntraeger (Fig. 25); Geologische Rundschau (Fig. 28); The Geological Society of London (Fig. 9); Keter Publishing House (Figs 55, 70, 107 and 108); New Science Publications (Fig. 61); Pergamon Press Ltd (Fig. 11); Reed Methuen (Fig. 49); The Royal Society of New Zealand (Figs 23 and 57); and The Seismological Society of America (Fig. 57). My apologies to those copyright holders we were unable to trace. Permission is not required to reproduce material published by the US Geological Survey, but I am grateful just the same.

October 1985   *C. V.-F.*

Chapter One

RECENT EARTH MOVEMENTS

. . . the very finger-touches of the last geological change.
Hardy, The Return of the Native

Scientific advances during the last three decades have made the study of recent earth movements increasingly effective. They have also given it theoretical and practical urgency. Novel surveying techniques, such as satellite laser ranging, allow us to measure displacements of the earth's surface which might otherwise go undetected; deformed strata and landforms can be dated by radiometric methods; exploration geophysics gives access to the changing sea floor; and dynamic crustal models provide a framework within which the search for change can fruitfully be conducted. The strongest demand for results comes from the very geophysicists and seismologists who not long ago revealed a restless globe of jostling continents, for without information on the nature and rate of the movements it is impossible to test and refine structural reconstructions and hypotheses; the civil engineer knows about the dangers created by crustal movements, especially sudden ones, and wishes to arm against them when dams, pipelines and nuclear power stations are at risk.

But one must not overstate the novelty of the topics. Quite apart from the work of convinced drifters, such as Du Toit's *Our Wandering Continents* (1937), books entitled *Our Mobile Earth* (Daly, 1926), *The Deformation of the Earth's Crust* (Bucher, 1933), *The Pulse of the Earth* (Umbgrove, 1942) and *The Unstable Earth* (Steers, 1945) antedate by three or four decades acceptance of plate tectonics and the agitation it implies. Indeed, earth movements figure prominently in the writings of Lyell, Darwin and their nineteenth-century contemporaries. In short, the shift from a 'stabilist' to a 'mobilist' set of beliefs over the last half century (Hallam, 1973,

1

1983) has legitimized what previously had seemed a somewhat eccentric set of beliefs.

Nevertheless the acceptance of mobility remains partial. *Seismicity of the Earth,* first published by Gutenberg and Richter in 1949, included a series of sketch maps depicting large tracts of the earth as stable (Fig. 1). Sixteen years later, the classic work on physical geology by Arthur Holmes (1965) referred to areas long immune from orogenic movements as stable shelves or platforms. Modern geology textbooks still classify the major units making up the earth's surface as either tectonically stable or tectonically unstable, and they apply the term craton to the supposedly stable central portions of continents (Ollier, 1981).

Stable is, of course, a word that can mean several things. The 1949 maps were drawn by seismologists, and to them stability conveyed freedom from earthquakes. Yet implicit in their picture is the contrast between ancient continental nuclei and mobile belts or geosynclines* in which prolonged sedimentation is in due course followed by compression and folding. Here stability equals rigidity or at any rate a lack of evidence to the contrary. In the words of Holmes, the stable shelves or platforms have not been subjected to orogenic (mountain-building) forces for hundreds of millions of years. The modern geophysicist will doubtless differentiate between various kinds of structural settings prone to instability, such as island arcs

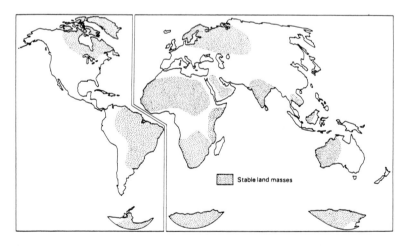

Figure 1. The world's stable areas according to Gutenberg and Richter (1949).

*For the contribution of J. Hall (1859) and J. D. Dana (1873) to the concept of geosynclines see Wyllie (1971) and Steers (1945).

and ocean ridges, or identify types of mantle that make for stability or instability in continental and oceanic areas (Wyllie, 1971). That is, he will be concerned chiefly with predictive models. Quaternary geologists and geomorphologists talk of stable coasts to mean those that have not shifted appreciably during the 2 million or so years that interest them. Thus Fennoscandia, parts of which have long been known to be rising by about 10 mm/year (Fig. 2), is viewed as a classic instance of isostatic readjustment by students of glaciation and mantle viscosity, yet it appears in many seismological reports as one of the archetypal stable shields.

It is not only the frame of reference that shifts from one observer to another. The field evidence is itself often open to more than one interpretation. Take the kind of data that are central to studies of Scandinavian uplift, namely the progressive seaward displacement of the coast. The eustatic theory of Edward Suess, which became firmly

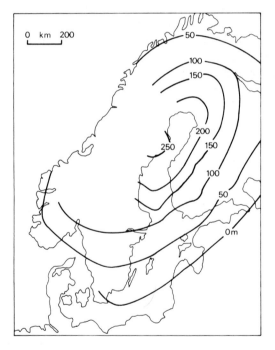

Figure 2. Uplift of the Baltic area since 6800 BC according to Sauramo (1939; after Zeuner, 1958). The isolines were based on the height of beach deposits of the first Rhabdonema Stage, which according to H. Thomasson postdated the Yoldia Sea whereas E. Nilsson later applied the name to the final stage of the Yoldia Sea, i.e. nearer 8000 BC, a fine example of the confusion created by the use of names for geological features.

established during the first quarter of the present century, saw the continents as fixed bastions round which the oceans lapped. There was a strong temptation to view shifts of the shorelines as largely due to fluctuations of sea level. As evidence for continental movements accumulated, the eustatic hypothesis began to creak (Chorley, 1963); but it refuses to croak. The latest challenge to it comes from studies of the geoid or sea-level surface, which is controlled by gravity and the earth's rotation and which displays topographic differences of almost 180 m (Fig. 3), for shifts in its configuration would greatly confuse the shoreline record (Mörner, 1980). But the eustatic model and the corollary of continental stability persist. Nowhere is this clearer than in petroleum geology, where fluctuations of sea level are perhaps more in vogue than

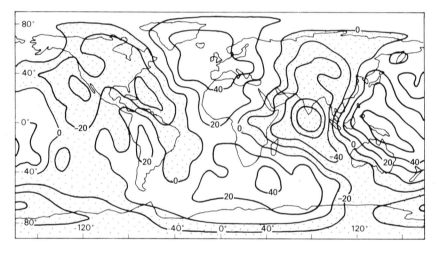

Figure 3. The geoid (Smithsonian Standard Earth II) relative to a spheroid of flattening 1:298·25. The lowest point is off southern India (– 113 m), the highest over New Guinea (+ 81 m). The map on which this figure is based was published in 1970. It was derived from simultaneous observations of satellites from several stations (more than 100 000 photographs were used for this purpose) and some of the first measurements by laser tracking of satellites. About 200 000 simultaneous equations had to be solved. The latest maps of the geoid reflect remarkable advances in laser tracking. The laser observations made in the late 1960s were accurate to a few metres; current techniques approach accuracies of 1 cm. Naturally there is a need for corresponding advances in gravity measurement. The GEM 8 model (1975) used 1600 surface gravity observations. It has produced slight changes: for example, the hump over New Guinea is now put at 74 m and the low off S. India at – 104 m. The latest systems use altimetry from the satellite itself. The above information and the figure are based on King-Hele (1971) with permission. ©1971 The American Association for the Advancement of Science.

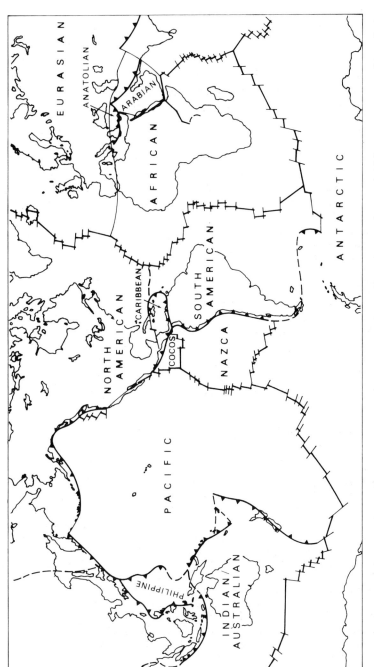

Figure 4. Major lithospheric plates. The heavy lines represent divergent boundaries, the toothed lines convergent boundaries, the light transverse lines transform faults. Based on Francheteau (1983) by permission of W. H. Freeman & Co.

ever for correlating stratigraphic sequences in different parts of the world and in seeking to explain the irregular distribution of oil and coal reserves (Vail *et al.*, 1977).

In any case the stable/unstable dualism is built into modern ideas on global structure. Tuzo Wilson has observed that, even when it becomes possible to demonstrate experimentally that the continents are moving relatively to each other, it will be impossible to demonstrate experimentally that such movement in the past accounts for cumulative displacements totalling hundreds of kilometres: acceptance of continental drift is more a question of mental attitude than of observation (J. T. Wilson, 1971, quoted by Miyashiro *et al.*, 1982, p. 68). And acceptance of the package implies belief in relatively rigid crustal units, originally called blocks (Morgan, 1968) and now known as plates, bordered or separated by belts where deformation of one kind or another is the rule (Fig. 4), a clear echo of the platform/geosyncline division. The idea can be traced back through

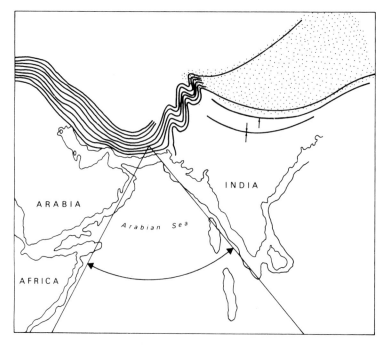

Figure 5. The orocline concept. Rotation of blocks as the sphenochasm opens can lead to thickening (stippled area) and the formation of an S-shaped orocline loop (thick folded symbol). The syncline in northern India is the Ganges trough and the line north of it is the Great Boundary thrust. The outer coastline represents the 2000 m isobath. After Carey (1976).

Holmes to the founding fathers of drift. In the 1920s, Emile Argand was advocating drastic horizontal displacements of portions of southern Europe and Asia as an alternative to the contractions of the earth's crust by which many important structural features were then explained (Argand, 1924). Wegener, whose book on continental drift was first published in 1915, viewed the continents as blocks which had been 'permanent throughout the earth's history' though not, of course, stationary or submerged by the same amount at all times. So do those, like S. W. Carey (1976), who accept drift but by a mechanism, in his case global expansion, which is not generally favoured (Fig. 5).

If some exponents of plate tectonics overemphasize the horizontal displacement of crustal units at the expense of vertical deformation, there is a comparable though contrary tendency among some of the critics of the plate model. Beloussov (1970), for example, has suggested that uplift originating in the mantle is the primary source of folding; Lyttleton (1982) favours global contraction as the key to mountain building; and Lester King (1983) believes that cymatogeny (uparching) has been the characteristic tectonic mechanism of the late Cainozoic (Fig. 6). Granted that some of the conflict stems from the nature of the data to hand, and some from political or other forms of churlishness, it is a caricature of the debate to present it thus polarized. Fold development may result from uplift by virtue of the tangential forces that the uplift can generate, and the many variants of the undation (wave development) theory likewise ascribe deformation to gravitational sliding and gliding of material from geotumours or swellings (e.g. van Bemmelen, 1954). More generally, uplift

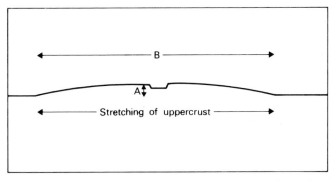

Figure 6. Diagrammatic section through a typical continental cymatogen showing major arching and rift. Valley at crest due to formation of tectonic gneisses at intermediate depths (not shown). A is 5000 ft (= 1520 m) or 'perhaps more'; B is 250 miles (= 400 km). The crux of the figure is that 'displacements are in the vertical sense only'. After L. C. King (1983) with permission. © 1983 John Wiley & Sons Ltd.

or subsidence are processes implicit in plate tectonics, witness the upthrust Tibetan Plateau or the extensional basins of the North Sea.

The study of recent earth movements cannot remain immune from such doctrinal disputes, and some will welcome the opportunity to collect data specifically for the purpose of testing, if necessary to destruction, competing hypotheses. But there are drawbacks to the 'anti-inductivist' line. It can lead to the neglect of sources which have been dismissed as uninteresting or unrewarding *a priori*. It also gives added impetus to the tendency for information to reach us 'through the strong filters of culture, hope, and expectation' (Gould, 1980, p. 118). Areas with a reputation for stability will tend to resist deformation. Features interpreted by Dr X as the product of coastal uplift are dismissed by Professor Y as eustatic. Fortunately most earth scientists combine herd instinct with a strong exploratory urge. Moreover many of those best placed to report recent earth movements are unaware of the hypotheses currently on the boil or dismiss discussions of scientific methodology as inconsequential.

As one would expect, earthquakes were being documented long before progressive uplift or depression attracted the attention of geologists. Quite apart from the loss of life and the destruction they bring, earthquakes are more easily recorded than are displacements which take decades or centuries to manifest themselves. But it does not follow that we are endowed with a long and faithful narrative of seismicity. A common complaint is that the search for an explanation has tended to get in the way of systematic observation. According to Richter (1958), 'Ancient accounts of earthquakes do not help us much; they are incomplete, and accuracy is usually sacrificed to make the most of a good story'. Useful reports do not appear until the eighteenth century (see Forbes, 1963) or, at the earliest with the publication of Robert Hooke's *Discourse on Earthquakes* in 1668 (Adams, 1938).

Seen from the viewpoint of the present book, Richter's assessment is reinforced by the tendency of many early observers to say little or nothing about any ground movements accompanying the earthquake and instead to describe the sounds or curious atmospheric phenomena that came to be associated with earthquakes. Adams shows how the views held on the subject in classical times laid the blame on one or other of the four elements. Thus Thales favoured water, Anaxagoras fire, Anaximenes earth and Archelaus air. Lucretius and Seneca also opted for air as the driving force. Vapours or winds continued to be favoured in this connexion in Europe throughout the Middle Ages and are still central to some popular explanations.

Mediaeval geology in Islam and China was, in the words of Joseph Needham, much less fanciful. The first seismograph, or at any rate a device which recorded the cardinal point closest to the direction of first ground motion (or seismoscope), was invented by Chang Heng in AD 132, and lists of earthquakes in China go back to 780 BC, but the explanation of earthquakes in that country was hampered by astrological, philosophical and political considerations (Needham, 1959, p. 624). The origin of mountains was handled more successfully. Chu Hsi, for example, writing in about AD 1170, remarks that 'the frontiers of sea and land are always changing and moving, mountains suddenly arise and rivers are sunk and drowned . . .'. The rest of the passage shows that Chu Hsi understood that uplift had taken place after the lifetime of the shells now entombed in the rocks, as Leonardo was to do three centuries later. Avicenna's Book of the Remedy, written in Arabic in Persia between AD 1021 and 1023, discussed sudden land uplift during earthquakes, although—like Aristotle before him—Avicenna ascribed the uplift to the winds that had produced the earthquake in the first place (Adams, 1938).

The notion of seismic uplift (Frontispiece) recurs in the later European literature. Agricola (Georg Bauer, 1494-1555), the 'Father of Mineralogy', listed earthquakes among the five mechanisms by which mountains could be produced. In 1561 Valerio Faenzi published *De Montium Origine*, in which he stated that earthquakes could develop islands as well as mountains. Robert Hooke (1635-1703) identified four possible effects produced by earthquakes. In terminology worthy of Lewis Carroll, he listed, in addition to the elevation and sinking of land areas and parts of the sea floor, 'Subversions, Conversions and Transpositions' and, finally '*Liquefaction, Baking, Calcining, Petrifaction, Transformation, Sublimation, Distillation,* etc.' (Hooke, 1705). Likewise there are many allusions to the rapid growth of volcanoes on land or beneath the sea to form islands, for instance in the writings of Anton Lazzaro Moro (1687-1740). Neither process was incompatible with the biblical chronology or with the common belief in a terrestrial history composed of a sequence of catastrophes.

More intriguing are the glimpses one is given of critical minds at work observing crustal processes of a more progressive character. Thus Faenzi noted that not all earthquakes were accompanied by fracturing of the ground, and he claimed that people had witnessed mountain building in progress. In 1734 Marie-Pompée Colonne wrote that mountains grew like plants, albeit at rates too slow for us to perceive. Steno (the Dane Niels Steensen, 1638-1686) drew attention to the effects of faulting on the topography of Tuscany and asserted that not all the mountains now in existence were present 'from the beginning of things'.

The benefits of field observation, manifest in the work of Steno, were soon to reveal the gradual elevation of Scandinavia. Convinced that sea level could not fall, Leopold von Buch (1774-1852) concluded from the landward displacement of marks made on the shore by Celsius and Linnaeus

a

May 1982

b

July 1983

Figure 7 (a)-(d) (above and opposite). The 'Temple of Serapis' (market) at Pozzuoli, Naples, in 1982, 1983 and 1984. Note the dark band on the columns produced by molluscan borings and the progressive emergence over two years. Courtesy of Dr J. E. Guest.

c

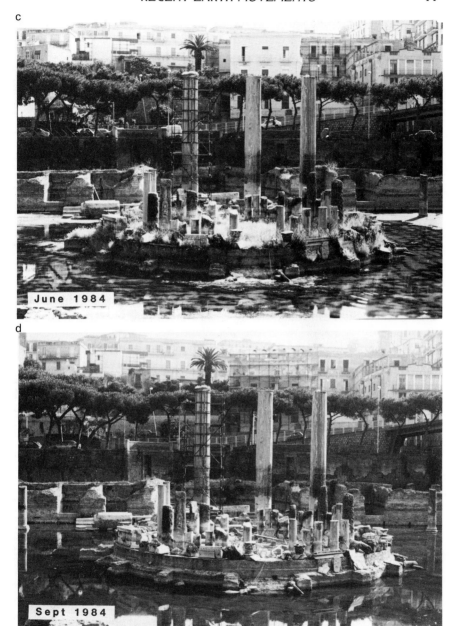

June 1984

d

Sept 1984

that both Sweden and Norway were being gradually uplifted [von Buch (1867) translated in Mather and Mason (1939)]. Note that Celsius had ascribed emergence to a 'diminution' of the ocean; a century later John Playfair recognized it as the product of aseismic uplift; yet in Charles Lyell's *Principles of Geology* (1830-33) the issue was still presented as a controversial one (see 5th edn, Lyell 1837).

Lyell's caution may in this instance strike us as excessive; yet there are many locations (and Scandinavia is by no means finally excluded from the list) where premature explanation has jeopardized the quality of the field data. The success of the *Principles* in swinging the majority in favour of a uniformitarian view of geological history owed something to its emphasis on careful fact-gathering guided, to be sure, by some working hypothesis but not to the point of prejudice. After all, uniformitarianism was chiefly an antidote to obscurantism in its appeal to mechanisms which were known to exist rather than devices which could only be imagined. It could thus accommodate modest catastrophes, including earthquakes and volcanic eruptions, and the *Principles* contains numerous examples of the varied topographic changes known to have accompanied historical earthquakes. But even changes whose explanation was uncertain could be of value provided they were well documented.

The 'Temple of Serapis' at Pozzuoli illustrates this category to perfection (Fig. 7). The first full account of the ancient market place was by Charles Babbage, the mathematician. Six years after visiting the site, Babbage, (1847; the paper was first read in 1834) was able to show, after careful analysis of the local geology and the molluscan borings in three columns still left standing, that the ruins had subsided into the sea and then been gradually uplifted. He readily disposed of the suggestion that the evidence could be explained by the creation of a lagoon above sea level behind a bar thrown up by a storm. Babbage's own explanation for the movements was that the rocks beneath the pillars had expanded under the action of heat and contracted on cooling. Lyell, who used the pillars for the frontispiece of the tenth and later editions of his *Principles*, initially thought they formed parts of baths built below sea level and subsequently uplifted, but in time he came to interpret them as a temple which had undergone subsidence followed by emergence (Bailey, 1962, p. 112; Wilson, 1970). There is still uncertainty over the source of the movements but no one disputes that the site demonstrates rapid displacement first downwards and then upwards. Compare the case of Paestum, where a precipitate search for evidence of submergence prompted the observation, later proved false, that there too the marks of intertidal borings could be seen on ancient pillars (Flemming, 1969).

Babbage's survey bears the stamp of what might be regarded as present-day standards of field investigation. Darwin's observations on coastal uplift in South America have likewise not been superseded in their essentials by more recent studies. The survey of the fossil shorelines of Lake Bonneville, carried out by G. K. Gilbert in 1881, has proved astonishingly durable. Better field equipment and decades of scientific progress do not invariably triumph (see Chapter 2). Where technology has spelt progress in the last century is in earthquake research, as both fixed and portable seismographs have transformed the subject. Yet the importance of astute field inspection of deformed areas remains undiluted. The memoir written by the Japanese seismologist Bunjiro Koto on the Kumamoto earthquake of 1889 placed the reality of major crustal displacements 'quite beyond dispute' (Davison, 1937). R. D. Oldham threw light on many branches of seismology; for the newcomer to the study of seismic deformation, his field descriptions following the 1897 Indian earthquake remain exemplary not least in showing what can be accomplished by a single, determined observer in difficult terrain (Oldham, 1899, 1926).

NEOTECTONICS

In 1962 Goguel, the author of a standard text on tectonics, remarked:

> The doctrine of uniformitarianism, so essential in geology, generally implies the preliminary study of present phenomena, which should enlighten us regarding the role of past phenomena, which we are seeking to reconstruct. This method hardly applies to the study of tectonic deformations, since the number of cases in which we have been able to demonstrate a movement directly is very limited (Goguel, 1962, p. 14).

Nevertheless research into the subject was being actively pursued in many parts of the world. Certain problems, such as the uplift of Scandinavia and kindred phenomena in North America, continued to attract attention. The idea that the movements stemmed from isostatic rebound following deglaciation was first mooted in the late nineteenth century and gave added impetus to this work because it allowed the results to shed light on the physical character of the crust and mantle. Gilbert's studies of Lake Bonneville opened the door to isostatic interpretation of lake basins, deltas and other land areas which had undergone loading by sedimentation or unloading by erosion. Similarly, the study of ground deformation during earthquakes continued to advance, although it could still be observed in 1982 that 'remarkably few seismologists in Europe and the Middle East have had the opportunity of studying earthquake effects in the field' (Ambraseys and Melville, 1982, p. 171 n. 21).

By the middle of the present century the traditional lines of research on recent earth movements had been joined by a geodetic strand. Repeated first-order triangulation in continental areas, especially in central Europe, had revealed discrepancies too great and consistent to be explained by instrumental error, and a flourishing field of study was born. By its very nature, levelling is equipped best to reveal changes in elevation. In addition, some European geologists are inclined to believe that oscillatory vertical movements are the 'basic' form of geotectonic phenomenon (Beloussov, 1962). Even so, the data collected by the geodesists provide a useful counterweight to the emphasis on horizontal movements that comes from reliance on geological maps and remote sensing and that is fuelled by the search for interplate slip vectors.

The need remains to co-ordinate the various lines of enquiry. In 1954 a Special Study Group of the International Union of Geodesy and Geophysics was set up by Vening Meinesz for the 'Determination of Changes in the Earth's Crust in the Horizontal and Vertical Sense'. A permanent commission on the subject came into being in 1960 and several international symposia followed.

For many earth scientists the field has become identified with 'neotectonics'. The term, which was apparently coined by V. Obruchev (Gorshkov and Yakushova, 1967, p. 429), is defined in one geological dictionary as the study of rock deformation and dislocation during the late Tertiary and the Quaternary (Manzoni, 1968). A leading worker in that field has specified its scope in another way: 'the period within which one can extrapolate geophysical observations in the light of geological data' (Angelier, 1976). In 1978 an international commission devoted to the Holocene proposed the following definition for neotectonics: 'Any earth movements or deformations of the geodetic reference level, their mechanisms, their geological origin (however old they may be), their implications for various practical purposes and their future extrapolations' (*Striolae*, the newsletter of the International Association for Quaternary Research, 1982, vol. 4, p. 37). The failure to identify the period of study is defended on the grounds that, whereas some issues can be resolved with the 'instantaneous' data of seismology, others need access to geological chronologies which extend over the last 10^7 years.

In brief, neotectonics stands for the study of late Cainozoic deformation. Although the continuity between its subject matter and present-day movements lends it some distinctiveness, there is an unfortunate if implied contrast with some sort of 'palaeotectonics', not to mention a hint that, as in neo-classicism, a revival of interest has occurred. But the word is here to stay and it undoubtedly reflects widespread and lively interest in crustal movements which are recent enough to permit detailed analysis and the assessment of rates of change.

The development is to be welcomed, whatever the cost in neologisms, because measurement and chronology are crucial to the testing and improvement of models of the earth's interior. One is otherwise left with a ragbag of curiosities or at best a set of observations which can be made to fit the ruling hypothesis. Geodetic techniques are being employed to measure the spreading across the Icelandic rifts for which average rates have been computed from the volcanic and palaeomagnetic record. The crustal shortening produced by the northward drift of India also lends itself to direct assessment, this time by comparing the original and current position of markers set up by the surveyors who made the first trans-Himalyan links in 1880. This kind of research, which Wegener used as the starting point for his exposition of continental drift, is bound to gain impetus from the development of satellite technology. Radiocarbon dating applied to deformed beds has already permitted some test to be made of the shortening in coastal areas predicted from the oceanic record. And it is by the assessment of slip rates over millennia that seismic 'gaps' and aseismic creep can be evaluated as an adjunct to what is already an active area of research, the historical study of earthquakes.

Information bearing on recent earth movements is often subdivided into geological, geomorphological, historical and geodetic (e.g. Beloussov, 1962). The headings adopted in this book depart little from this scheme. Archaeological sources are included with those of history (Fig. 8), although some dates derived from artefacts are necessarily included in the geological sections. Geodetic data are considered together with eyewitness accounts on the grounds that both represent the direct observation of movement whereas movement can only be inferred from the evidence of geology, geomorphology and history. Likewise, instrumental measurements that depend on devices which record changes in the earth's form as they take place are accorded separate treatment. These distinctions are not regarded as rigid or scientifically profound. Their sole task is to facilitate discussion of a varied assemblage of reports and deductions.

The second part of the book is also arranged with convenience as the main consideration. Thus, seismological applications are discussed separately from other geophysical matters simply because years of collaborative data gathering by seismologists and geologists have brought into existence a branch of earth science—seismotectonics—in which the analysis of recent ground movements plays an integral part. Again, there is something to be gained from looking at our findings from the standpoint of a workaday geologist who wishes to reassess the ground rules of his subject, as it were, in the light of recent deformation. Finally, the needs and interests of the

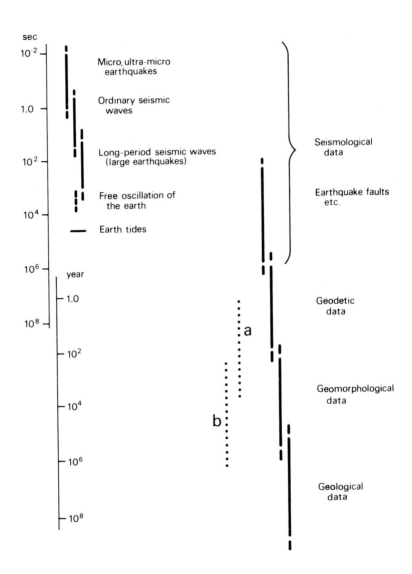

Figure 8. The right-hand column shows the time ranges covered by the major sources of evidence bearing on recent earth movements. Seismological and other vibratory data (left-hand column) are used mainly in interpreting the movements revealed by other sources. After Kasahara (1971, fig. 1.5). Historical and archaeological evidence have been added (a and b respectively).

engineer can usefully be considered without any prior assumptions about the triviality of ground movements in non-seismic regions. For, besides our defective knowledge of how far these regions extend, there is no justification for thinking that gradual movements are necessarily harmless or, indeed, that stability can ever be taken for granted.

A Note on Earthquake Magnitude

Some of the authors cited in the references do not specify which sort of magnitude they mean. When they do, I follow suit. On the whole the laxity does no harm if (as in this book) the arguments are very general. And, besides the uncertainties inherent in the calculations [$0 \cdot 2$-$0 \cdot 3$ according to Kasahara (1981)—and these are logarithmic scales)], Richter (1958) points out that, although M_L (Richter or local magnitude) is greater than M_b (or m_b = body wave magnitude, generally used for deep earthquakes) above $6 \cdot 75$, the two agree closely at magnitudes near $6 \cdot 75$. M_s, used for shallow earthquakes at epicentral distances of over 600 km, tends to bear a simple relationship to m_b. According to Bolt (1978), for example, it is approximately $m_b = 2 \cdot 5 + 0 \cdot 63 M_s$.

Chapter Two

GEOLOGICAL EVIDENCE

Scarcely any fact struck me more when examining many hundred miles of the South American coasts, which have been upraised several hundred feet within the recent period, than the absence of any recent deposits sufficiently extensive to last for even a short geological period.

Darwin, The Origin of Species

Like most records, the geological evidence from which earth movements are reconstructed is often ambiguous and always open to more than one interpretation. The three major items at issue are the nature of the deformation, its age and the mechanisms responsible for it. The first of these is just as likely to provoke disputes as the other two. As Darwin observed in the passage cited above, recent sediments tend to be ephemeral; and conjecture thrives on incomplete data. But that can hardly account for every controversy, especially as a superabundance of facts does not guarantee unanimity. A more convincing explanation is that, although field workers are no more prone to preconception than any other kind of investigator, field data are inherently ambiguous.

An excellent illustration of this state of affairs is provided by the Dead Sea depression (Fig. 9), which forms part of a major structure to which belong the Red Sea and the East African Rifts. According to one school, the field evidence can be explained by normal or reverse faulting, that is to say by vertical displacements. The opposing view is that horizontal movements have predominated, with the resulting drag held responsible for transverse folds and localized vertical faulting. Both interpretations have had their proponents for over a century, and one might be forgiven for thinking that the current supremacy of the horizontal school is now due to fade.

In 1896 M. Blanckenhorn suggested that the Dead Sea occupied a downdropped block (or graben) bounded by flexures and locally by normal

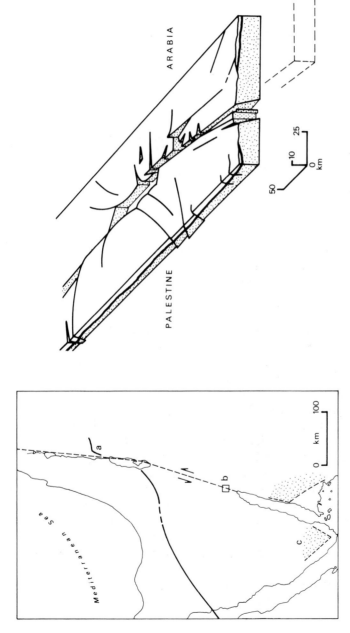

Figure 9. The Dead Sea Rift. Left: Evidence for left-lateral displacement includes (a) the northwestern boundary of Lower Turonian (Upper Cretaceous) ammonites and (c) the margin of PreCambrian blocks (stipple). Adapted from Freund (1965), with permission. About 150 m of left-lateral displacement in alluvial fans less than 20 000 years old is reported from (b) by Zak & Freund (1966). Right: Block diagram based on Quennell (1958) showing part of Arabia and Palestine blocks now and formerly (dashed line in foreground). Note how movement has led to compression where two protuberances have been juxtaposed by slip and has produced tension fractures in gaps produced by the movement.

faults. This view was to be held in various guises by later authors. In 1929 B. Willis described the trench as a ramp valley, that is a graben bordered— indeed, held down—by reverse faults [Fig. 10(1)]. In 1952 Lees explained the depression as a product of folding with the bounding faults (Fig. 11) as a by-product of shortening at depth. Normal faulting remains the dominant mechanism in the accounts of the area by Wetzel and Morton (1959), de Sitter (1962) and Bender (1974).

The idea that the Sinai-Palestine block has been displaced south relative to Transjordan can be traced back to Lartet in 1869. It was developed in 1932 by Dubertret, who compared the position of Cambrian, Triassic and Jurassic outcrops on opposite sides of the valley and concluded that the N-S displacement amounted to 160 km. An important contribution was made by Quennell in 1958 by combining geological (including structural) evidence with geomorphological observations. As he remarked in the introduction to his key paper, where geomorphic evidence for a horizontal component of movement is absent 'the dominance of the vertical component is often taken for granted. The classical interpretations of the faulting responsible for the topographic phenomena known as the "Rift Valleys" have been unduly influenced in this manner' (Quennell, 1958, p. 2; see also Burdon, 1959, pp. 56-65).

With the help of projected topographic profiles and reconstructed stream profiles, Quennell was able to identify elevated erosion surfaces from which he inferred phases of tilting, warping and local uplift and also changes in base level. These various episodes were ascribed to two main phases of horizontal movement on the wrench faults of the Rift. Where movement led to convergence there was down- or upwarping, whereas separation produced little distortion (Fig. 9). The first phase took place during the early Miocene-Pliocene, and amounted to 62 km. The second period dates from the late Pleistocene and is still in progress; it has already produced

---

Figure 10 (opposite page). Some definitions and problems in the field interpretation of faults. (1) from left to right: sinistral and dextral strike-slip faults; normal faults giving rise to a graben; reverse faults bordering a ramp valley. (2) Left: rotation across listric normal faults and secondary structures produced by accommodation faults, including a graben (g) and a horst (h). After Gibbs (1983), with permission, Pergamon Press. Right: normal fault system showing uplift of the footwall (f) as well as depression of the hanging wall (hw). Note antithetic faults (a). After Jackson and McKenzie (1983), with permission, Pergamon Press. (3) Vertical offset Y produced by strike-slip movement on a fault which cuts a slope. After Sieh (1978a), with permission from The American Geophysical Union. (4) Fault break which might erroneously be interpreted as a reverse fault. After King et al. (1981). (5) Strike-slip motion counterfeited by erosion of an area affected by normal faulting. After Hobbs et al. (1976).

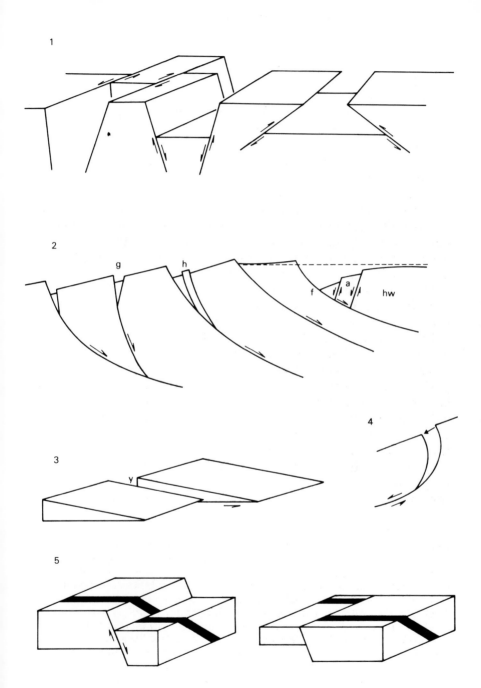

Figure 10.

Figure 11. Pleistocene Lisan Marls (right foreground) within Dead Sea Rift.

45 km of displacement. Freund (1965) initially put the total displacement since the Upper Cretaceous at 70–80 km but he later accepted a value of 105 km (Garfunkel *et al.*, 1981).

In a review of the pros and cons of the hypothesis, Burdon (1959) found much in its favour but drew attention, among other things, to the difficulty of accommodating over 100 km of displacement within the regional structure, and to the lack of local geomorphological evidence for the large proportion ascribed to the late Pleistocene. Zak and Freund (1966) then found traces of 150 m of sinistral strike-slip movement in the southern rift in alluvial deposits thought to be less than 20 000 years old. The resulting average of over 7·5 m/1000 years is of the right order of magnitude to account for the requisite 45 km of Pleistocene movement. If the deposits correspond to the Older Fill alluvial series of the Mediterranean Basin (Vita-Finzi, 1969), whose deposition stopped some 10 000 years ago, the rate (15 m/1000 yr) is in even better agreement with the long-term average. Such field observations have undoubtedly favoured the wrench model. But the critical impetus has come from the swing in the climate of geological opinion in favour of strike-slip movement between crustal blocks coupled with independent evidence for Eocene and later opening of the Gulf of Aden and the Red Sea. For extensive shear along the Jordan valley is wholly consistent with the anticlockwise rotation and northward shift of Arabia implicit in the sea-floor record and with the shortening indicated by the Zagros folds.

Throughout, one is reminded of the reaction of F. R. S. Henson to Quennell's 1958 paper: local evidence of tension, compression and horizontal shearing 'can be found to support any of the preferred hypotheses (rift-valley, ramp-valley, wrench-fault) which might explain Dead Sea tectonics' (in Quennell, 1958, pp. 19–20). Perhaps some kinds of field evidence are destined to remain ambiguous.

FAULTS

On air photographs or satellite imagery any prominent feature which is abnormally straight or gently curved and which separates contrasting terrains is commonly suspected of being a fault. Indeed, once they leave the ground, geologists and geophysicists who are usually models of scientific restraint will draw lines and interpret them as faults of a particular kind with startling fluency. The alarm among conservative geomorphologists who have been trained to distrust anything so infantile as linearity when seeking fractures is heightened by the realization that the method often works. In western Turkey, for example the surface breaks of the larger earthquakes and the seismic evidence confirm the impression obtained from

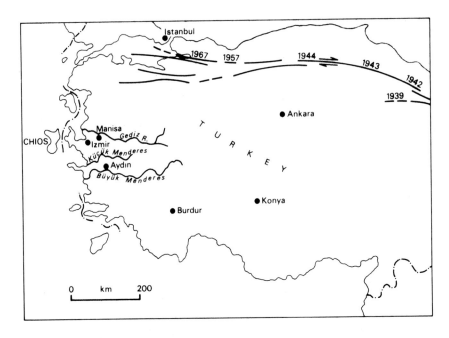

Figure 12. Western Anatolia. Only part of the North Anatolian Fault zone is shown. The dates are near parts of the zone where faults were activated during specific earthquakes (see also Jackson and McKenzie, 1984). The pattern illustrates the need to complement geological data with historical evidence when evaluating fault activity. After Hancock and Barka (1980) with permission.
© 1980 Macmillan Journals Ltd.

LANDSAT photographs that, south of the North Anatolian Fault (Fig. 12), current deformation is dominated by normal faulting (Allen, 1975; McKenzie, 1978).

Field checking is needed to show whether the lineaments are genuine faults rather than mental constructs. The reverse fault shown in Fig. 13 was found in southern Iran by the author and two colleagues after they had followed one such lineament for several kilometres across terrain which lacked any obvious sign of faulting when viewed at close quarters, and it demonstrates how the disturbance produced by faulting can be reflected in slight variations of drainage and vegetation even when the direct topographic effect is trivial. Of course, one cannot assume that the observed displacement applies throughout the inferred fault trace but, provided the assumption is treated as provisional, its adoption may prove helpful in planning the next stage in the fieldwork.

All too often, however, the nature of the faulting may be wholly obscured

Figure 13. Reverse fault affecting Neogene beds and overlying Late Quaternary gravels at Teleng, Iran. The fault dips 30° NW and strikes N 20° E; the northwestern block has been thrust over the southeastern by 5·5 m.

Figure 14. Chah Shirin fault, southern Iran, in Neogene beds and gravels laid down c. 30 000-7000 years ago. As a second set of deposits dating from 1250-300 years ago is undisturbed, faulting dates from 7000-1250 years ago. Note the dip in the skyline where the fault outcrops.

by erosion or poor exposure. Normal faults can then produce the illusion of strike-slip movement [Fig. 11(5)]. The dip of a fault may also prove surprisingly elusive. Variations from place to place along the strike are identified by multiple measurements. Changes in dip too deep to be exposed in section will only emerge from geophysical sources or drilling. In the absence of such data, fault planes are generally drawn concave upward rather than straight. In areas which have recently undergone extension this is explained by the current view that many normal faults are listric [Fig. 11(2)] and, as crustal shortening is sometimes preceded by stretching, a concave-upward form may be retained by the normal faults when they are reactivated as reverse faults. But faults which start as thrusts can also be of a listric type; again, concavity can arise from the need to accommodate rotations and from the greater rotation that is possible in the deeper layers in the crust where movement is by creep rather than brittle failure (Gibbs, 1983; Jackson and McKenzie, 1983).

The rotation of fault blocks is inevitable in listric fault movement. Together with accommodation faults, whether parallel to (synthetic) or at a sizable angle with (antithetic) the initial fault, the rotation can produce reverse faulting, horsts and grabens in various combinations. It can thus create further misleading data on slickensides as well as confusing topographic evidence.

The stratagems usually prescribed for the identification of faults depend on the omission or repetition of strata or the presence of cataclastic zones along which the country rock has been sheared, and therefore remain of value only when the rupture is in consolidated rocks with suitable stratification or texture. When the faults cut unconsolidated and perhaps poorly bedded or monotonous clastics, one may have to rely on morphology for identifying the fault and use conventional methods to reinforce the inference. The Chah Shirin Fault north of Bandar Abbas, for example, is visible on air photographs and emerges at one point as a reverse fault (Fig. 14). To trace it one relies on dips in the skyline, changes in the vegetation pattern and other such clues, and at some points the only confirmation comes from the rotation which some of the pebbles close to the fault plane have evidently undergone.

The problems encountered when formulating a chronology of faulting are equally knotty. If offset channels or beds are present (Figs 15 and 16) their age provides a maximum for the displacement and says nothing about its rhythm. Inadequate dating is also inimical to resolving such matters as the synchrony of fault episodes at separate locations or the relationship of the faulting to other processes, such as uplift or erosional unloading.

A few reliable numerical ages are worth a library of correlations based on assumptions. Because radiocarbon (^{14}C) dating has been available for

Figure 15. View northwestwards across San Andreas Fault in the Carrizo Plain showing stream channels dextrally offset. The one in the centre has been offset about 130 m during the last 3700 years. After Wallace and Schulz (1983) and Sieh and Jahns (1984).

little over 30 years it has produced a number of unreliable or wrong dates. Opponents of all ^{14}C dating have welcomed such errors as evidence that the method does not work but all it proves is that people can be careless. Errors are not unknown in the other main sources of quantitative ages, namely historical research, man-made structures and eyewitness reports. Deposits and landforms too old for ^{14}C dating may yield material suitable for other radiometric methods, such as volcanic deposits for potassium/argon (K/Ar) dating or carbonates for U-series dating.

If no ages are forthcoming, one will of course have to make do with ordinary stratigraphic methods though preferably calibrated by radiometric dating elsewhere. Relative dating is of little value to the main applications of fault chronology: calculating recurrence intervals (or repeat times) for earthquakes, and comparing the deformation predicted by geophysical and structural models with what actually happened.

Radiocarbon dating of shelly sediments has made it possible to trace progressive vertical offsets on the Coyote Creek Fault, part of the San Jacinto Fault zone (Fig. 17). A section of the fault ruptured during the Borrego Mountain earthquake of 1968; the offsets suggested that events of similar magnitude (6·4) had occurred every 160 or 190 years during the last 860 years. Once warping as well as offsets were taken into account, the recurrence interval came out as 205 years for the last 3080 years; vertical offsets alone indicated 195 years for the last 1230 years. An interval of

Figure 16. Normal fault on Konarak Peninsula, Iran (the fault runs diagonally towards lower left) capped by undisturbed Quaternary marine deposits. The scale is indicated by the human figure.

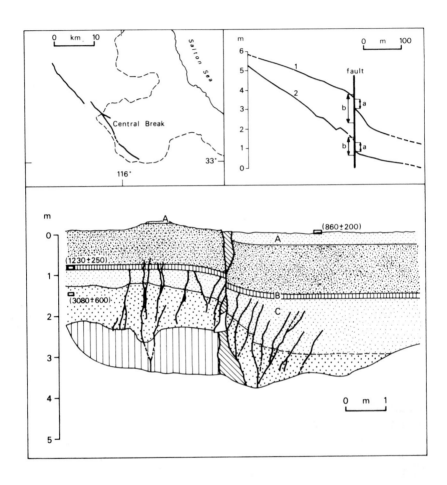

Figure 17. The Borrego Mountain earthquake of 1968. Top left: the three principal breaks in relation to the highest shoreline of Holocene Lake Cahuilla. Top right: two SW-NE profiles across the scarp of the central break showing 450 mm of fault slip (a) ascribed to the last 860 years, and the total tectonic offset (b) indicated by displacement of the base of the youngest lake bed (viz. layer A laid down about 860 years ago). The discrepancy suggests that there has been drag as well as fault slip. Lower: profile exposed by a trench across a break a few metres northeast of the central break showing that the older the strata the greater their displacement. Thus the base of layer A is 560 mm lower to the right (west) of the main fault than it is to the left; for layer B the displacement amounts to 740 mm, and for layer C 1·7 m. The boxes and numbers refer to radiocarbon dates. After Clark et al. (1972).

about 200 years is thus indicated, and it is compatible both with the historical evidence of the San Jacinto zone and the long-term values derived for the San Andreas Fault north of the Transverse Hill Ranges (Clark *et al.*, 1972).

The shelly beds had been laid down in a freshwater lake which occupied the Salton Basin in the late Holocene. The youngest deposit laid down in the lake in what is now the Lower Borrego valley is about 860 years old; the oldest ^{14}C age in the section shown in Fig. 17 (W-2468) is on gastropod shells within a bed locally over 1 m thick. If the stratigraphic boundary above this sample were to prove significantly younger than 3080 years, the recurrence interval would be less than 205 years.

Besides this kind of field uncertainty, there is the possibility that the shells were not in isotopic equilibrium with atmospheric carbon, a problem that has to be faced whenever non-marine molluscs are used. Nevertheless, one of the species in question, the mussel *Anodonta*, gave dates elsewhere in Lake Cahuilla which agreed well with results obtained with charcoal and tufa, and a constant rate is indicated by the graph of offset against age.

The geological study of fault history also stands to benefit from evidence linking the displacements to earthquakes. In the present climate of opinion the temptation is strong to assume that movement on faults is seismic, whereas measurements sometimes show that creep has predominated. Even if earthquakes are demonstrably to blame, some idea of their frequency is desirable.

Outside the USA, and indeed in less favoured parts of it, trenching is either forbidden or prohibitively expensive and reliance has to be placed on natural exposures, with the consequence that the sequence is generally incomplete. Radiocarbon dating is restricted for the same reasons. The development of first-order ^{14}C-dating methods, already practicable for Holocene shell carbonate, should help to reduce the financial constraints (Vita-Finzi, 1983), just as the accelerator has already made it possible to date strata containing only a few milligrams of carbon. Lightweight geophysical prospecting may prove equally helpful in locating sections worthy of excavation, and progress in the dating of archaeological assemblages will similarly benefit the geologist forced to rely on artefacts for establishing the age of land surfaces or deposits.

Reconnaissance studies of value can be conducted even when there is no datable material to be recovered. For instance, under favourable circumstances an estimate of the age of fault scarps can be derived from their height and slope (Wallace, 1977). In part of the Basin and Range province of the USA it was found that the principal slope of young fault scarps declined with age so that scarps which were about 12 000 years old had maximum angles of 20–25° whereas scarps which were much older could have slopes of 8-9°. What is more, repeated displacement on a fault produced composite or multiple scarps.

Naturally such generalizations can be made only if the comparison is made between features of similar height in similar rock types and under comparable climatic conditions. In one such study, conducted in western Utah, it was found that relative age could thus be assigned to scarps in alluvium over a time range between several thousand and several hundred thousand years (Bucknam and Anderson, 1978). At the very least this

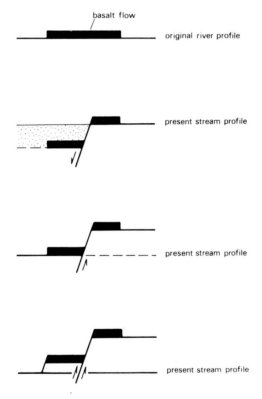

Figure 18. Stream profiles preserved by lava flows as marker beds. Note how the relationship of the topography to the original stream profile makes it possible to identify uplift of the footwall (left), the hanging wall (right) or both. In the first case, the channel may adjust by deposition on the hanging wall. In the second, it is likely that the profile will be below its original level on the foot-wall side. In the third, there will be erosion on both blocks. After Hamblin (1984), who used this kind of clue to identify the direction of absolute movement along the eastern boundary faults of the Basin and Range province of the USA. Reproduced with permission of The Geological Society of America.

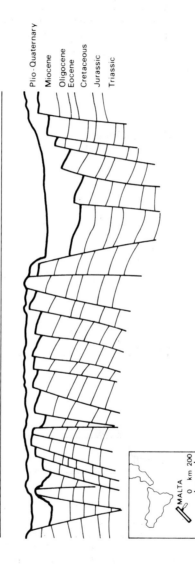

SW

MALTA GRABEN

NE

Plio-Quaternary
Miocene
Oligocene
Eocene
Cretaceous
Jurassic
Triassic

MALTA

0   km   200

Figure 19. Horsts and grabens south of Sicily: stratigraphic and structural interpretation of seismic line MS-19. If the stratigraphic interpretation is correct, the faults became inactive by the time the Plio-Quaternary deposits were laid down. Based on Finetti and Morelli (1972) by permission.

approach permits rapid age mapping from topographic data in preliminary analysis of regional fault history. We can expect further such productive ideas from field workers impatient with the reluctance among their colleagues to draw generalizations despite decades of observation.

A combination of geomorphological analysis and radiometric dating may help to resolve ambiguities in the geometry of motion as well as the issue of chronology. In Utah and Arizona this has been done by comparing modern stream profiles with their predecessors fossilized beneath Cainozoic basalt flows. The comparison may reveal not only the nature of the fault but also whether it was the hanging- rather than the foot-wall block that was primarily responsible for the displacement (Fig. 18). In the example illustrated, with uplift of the foot wall downcutting is promoted, whereas if the hanging wall is depressed the outcome is likely to be aggradation. There are parts of the Dead Sea basin where this simple yet powerful approach (Hamblin, 1984) could perhaps extract novel information from the river and topographic profiles discussed earlier in the chapter. It is also probably the case that in the absence of lava flows many faults in unconsolidated sediments are rapidly obliterated or go undetected (Huntoon, 1977).

Submarine faults present special problems. Given that they have been identified correctly from the geophysical data (Fig. 19), their dating relies on correlation with cores for which radiometric ages are available, but all too often coring and dating are confined to parts of the sea floor which are little disturbed by faulting, because it is there that a continuous depositional record is likely to be obtained. As a result the chronology of deformation may need to be expressed on a time scale which is only crudely (and subjectively) subdivided. For instance, a 'sparker' survey of the Pelagian shelf off eastern Tunisia and northwestern Libya revealed three units (Bellaiche and Blanpied, 1979). The lowest, which includes a group of strong reflectors, is thought to date from the late Miocene. The middle one, which is relatively transparent, is ascribed to the Lower Pliocene. The uppermost contains many well stratified reflectors and is equated with the Upper Pliocene and the Quaternary. A disconformity which can be recognized in some places is taken to represent the base of the Quaternary partly on the grounds that in parts of Tunisia the close of the Neogene is characterized by an erosion surface.

Drilling alone can supply the requisite ages: dredging from canyons and fault scarps is unlikely, even with the help of submersibles, to match in importance the data yielded by natural or artificial sections on land. Where the sea scores over the land is in the accumulation of volcanic ashes which can be dated radiometrically and traced over large distances. Likewise, palaeomagnetic correlation is often possible between cores taken from the sea floor, whereupon the subdivisions obtained from geophysical records can be checked and broadly dated.

WARPING

One might expect a section on faults to be followed by one devoted to folds, but a more general heading is used in order to include tilting and other modes of deformation which are too amorphous or on too large a scale to qualify as folding in the familiar sense.

The deformation of strata and topographic surfaces is likely to be more difficult to observe than offset beds and fault scarps unless deposition or erosion was linked in some way to a water level. To be sure, depositional dip or the seaward slope of wave-cut platforms leads such features to depart from the horizontal even when they are fresh. But if the feature is extensive or deformed enough, the advantages are soon clear. The fossil beaches of Lake Bonneville, first studied in 1890 (see below, p. 145), were seen to depart from the horizontal despite the simplicity of the surveying techniques employed (Gilbert, 1890).

Simple seaward tilting has been inferred from a series of wave-cut platforms in central California because they are steeper than the modern

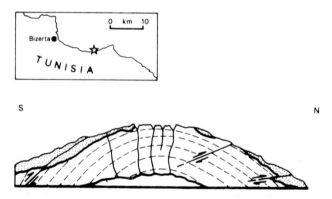

Figure 20. Folded Quaternary marine deposits at R'mel, in northern Tunisia, viewed from the east. The simplified section, based on Ben Ayed et al. (1978), shows Tyrrhenian (Quaternary) marine beds (light stipple) with dips of less than 20° and the fold axis striking 85° resting unconformably on an asymmetric fold in Pliocene rocks with dips of 40° on the southern limb and 25° on the northern and the fold axis striking 80°. Note the reverse faults which affect both units. The Tyrrhenian beds in the area are generally ascribed to the period between 124 000 and 80 000 years ago (Paskoff and Sanlaville, 1980). Folding occurred before, during and after this period. See Figs 21 and 22.

platform and the discrepancy is greater the older the feature (Bradley and Griggs, 1976). The youngest terrace affected by the process is thought to be 125 000 years old; anything more recent has been destroyed by erosion, including wave action when sea level was higher than the present. The implication would seem to be that, even in a zone as tectonically mobile as coastal California, the survival of a full record requires the right component of net uplift.

Figure 21. Late Quaternary marine deposits folded into an anticline at R'mel, Tunisia.

The growth of coastal anticlines has been investigated in various areas with deformed Pleistocene beds. At R'mel, east of Tunis, an anticline in Pliocene beds is overlain discordantly by Pleistocene marine beds (Figs 20 and 21) which are also folded. Initial folding produced by N-S compression is thought to have occurred between the Pliocene and the Eutyrrhenian (*c.* 124 000 years ago). Renewed compression occurred during the Eutyrrhenian and Neotyrrhenian (*c.* 80 000 years ago). Finally, post-Neotyrrhenian compression gave rise to reverse faults and some tilting (Ben Ayed *et al.*, 1978). Yet it is conceivable that the feature is wholly of Late Pleistocene age. A sample of *Glycymeris glycymeris* collected from the Eutyrrhenian beds (Fig. 22) by Dr G. Richards and the author in 1981 was subjected to various tests for contamination before being submitted to Dr H. Polach for ^{14}C assay. The result conflicts with existing estimates,

Figure 22. Fossil molluscs within shallow-water and beach deposits at R'mel. Radiocarbon dating of a specimen of Glycymeris glycymeris from the beach has given an age of 33 160 ± 1400 years BP (ANU-2799); folding presumably occurred subsequently.

but the faunal assemblages by which Mediterranean beaches have generally been dated are of dubious chronological value (Richards, 1982).

The R'mel anticline is thought to reflect three phases of folding; But it could be that the discontinuities are largely the vagaries of erosion and deposition. Discontinuous uplift is by no means incompatible with fold development, as later sections on El Asnam (Algeria) and Tujak (Iran) should show.

In New Zealand, a fully three-dimensional picture of coastal synclines and anticlines affected by recent folding has been obtained by combining uplift data with estimates of progressive changes in the angle of tilt of the folded beds and of marine terraces cut across them (Lewis, 1971; Wellman, 1971). This approach makes it possible to detect changes in the plunge of an anticline as well as in its cross-sectional form. The onset of folding is given, of course, by the age of the youngest bed affected. At Hawke's Bay, folding has apparently occurred since the early Pliocene. The zero isobase separating land undergoing uplift from land subject to subsidence is slightly offshore (Fig. 23).

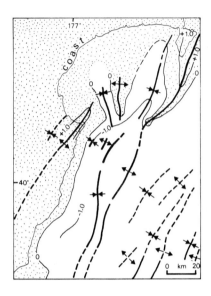

Figure 23. Hawke's Bay Land District, North Island, New Zealand. The stippled area is being uplifted; the fine lines indicate vertical movement in metres per thousand years. The anticlines (arrows diverging) and synclines (arrows converging) are thought to be growing. Those offshore have been inferred from seismic profiles or the bathymetry. After Lewis (1971).

Away from reference shorelines (Fig. 24), deformation can be detected only if the original form of the unit or topography in question is well established. River terraces are suited to this task if their form and gradient away from the affected area are exceptionally regular. The mere presence of river terraces standing above a river channel does not, of course, always signify tilting or warping. But, conversely, the tendency in recent years has been to reject tectonic factors in favour of explanations requiring climatic changes or human influence. The correct answer is doubtless a blend of several mechanisms.

Deposits which were laid down mainly by gravity, rather than in running or standing water, rarely produce bedding or topography whose original geometry is self-evident. Nevertheless tilting and warping may become apparent either because the present gradient or shape departs grossly from any modern examples or because cumulative change can be identified by working back from the latest item in a sequence. The Bakhtiyari conglomerates of Iran, which commonly dip steeply enough to indicate postdepositional tilting, have come to be regarded as the products of

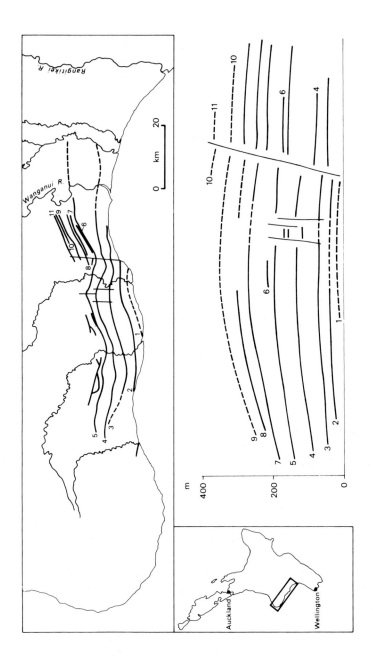

Figure 24. Deformed shorelines and faults in the Wanganui area, New Zealand. After Pillans (1983) with permission of The Geological Society of America.

Mio-Pliocene uplift along the Zagros on the grounds that mountain growth would not only deform the beds but also supply the gravel required for their development. The major drawback to the argument is that, as the deposits lack fossils, all deformed conglomerates postdating the Mio-Pliocene Agha Jari Formation tend to be ascribed to the Bakhtiyari unit. Moreover, though officially defined as 'late Pliocene or younger in age', the Bakhtiyari Formation has come to be viewed as the product of Pliocene movements alone (James and Wynd, 1965; Falcon, 1974). One can only hope that the gravels will eventually yield datable material. Beds ponded within the gravels are potential sources of molluscs and other organic material; they would be of added value for assessing the amount of tilt since they accumulated as, unlike the gravels, they would have been laid down horizontally.

REGIONAL DEFORMATION

The most familiar instances of regional deformation in the late Cainozoic are provided by northern North America and Fennoscandia. In the latter, at least, the isostatic explanation of postglacial uplift is not universally accepted, but here too the debate has been damped by deficiencies in the

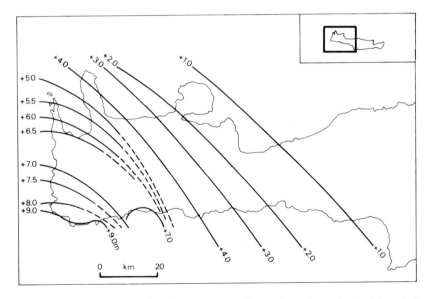

Figure 25. Holocene uplift (m) in western Crete, based on the height of the highest Holocene marine sea-level indicators. After Thommeret et al. *(1981).*

Table 1
Uplift (in mm/year) in glaciated areas.

Period (10³ years BC)	Ancient ice sheets						Present-day ice sheets and ice caps			
	Laurentide		European		Barents Sea		Greenland		Franz Josef Land	Severnaya Zemlja
	Centre	East periphery	Fennoscandian centre	Scottish centre	Spitzbergen	Kolgver Is.	East periphery	West periphery	Alexandra Land	Bolshevik Is.
10-8	100-70	70	30	10	13	18	60-70	30	8-10	12
4-3	30	3-5	10-15	4-5	3	1-2	5	5	1·5-2·5	

After Nikonov (1980). Figures (mm/year) corrected for eustatic sea level. The values for the Barents Sea are very tentative.

data. Detailed mapping at Glen Roy, in Scotland, has demonstrated dislocation of shorelines dated to 9600 and 6800 years ago and thus the action of abrupt, localised uplift in an area previously regarded as tectonically stable (Sissons and Cornish, 1982). The same applies to areas such as Crete (Fig. 25): recent uplift has long been depicted by smooth isolines marking a cumulative trend where detailed analysis supported by 60 ^{14}C dates indicates block movements of spasmodic character and with at least three reversals in their general trend (Thommeret *et al.*, 1981; Pirazzoli *et al.*, 1982).

The Holocene (Table 1) shows pronounced changes in the rate at which movement occurred in different glaciated regions. Variations of this kind are no less important than totals in elucidating the rheological properties of the crust and mantle, especially if the data are sufficiently detailed to reveal associated shifts in the pattern of deformation.

The usual difficulties faced in tracing folds and tilts arise anew when regional effects are being sought. A southeastward tilt of Britain is suggested by the displacement of coastal deposits laid down 6500 years ago and is consistent with the general stratigraphic record in that the oldest rocks are to be found in the northwest of the islands. However, if the original eustatic correction is revised in the light of current ideas, the submergence is transformed into stability or even emergence (see Chapter 9). Less familiar techniques for tracing uplift include the dating of cave stalagmites and related deposits (speleothems) which could accumulate only after the water table had been lowered—presumably by uplift—sufficiently for subterranean channels in limestone to become caves (Williams, 1982). Erosion rates, whether measured directly or derived from indirect measures of uplift and erosion—such as fission-trace ages (Wang *et al.*, 1982; Zeitler, 1982)—are evidently of interest both as rough pointers to the progress of uplift and because they may help to explain it (Jeffreys, 1970).

The temptation remains strong to infer tectonic events from geological evidence for which there are alternative explanations dependent upon different mechanisms according to the local circumstances. In parts of the northwestern Himalayas, thrusting raised local base-level and enhanced deposition in late basins; contrary to initial impressions, then, a lacustrine interval lasting almost 2 million years here reflects continued tectonism rather than quiescence (Burbank and Johnson, 1982). In the Caucasus the general pattern of deformation is inferred from drainage patterns, slope maps, variations in sediment thickness, shifts in the position of rivers and a host of other geological and geomorphological clues which are individually weak but persuasive in unison (Gerasimov, 1967; Trifonov, 1978). A similarly eclectic approach was followed some years ago in a concerted approach to date the uplift of the Andes in northern Chile (Hollingworth,

1964). Analysis of regional deformation on the scale of lithospheric plates will long continue to require the broad-brush results yielded by seismic profiling coupled with morphological analysis (Weissel *et al.*, 1980). Some of the most challenging facts of neotectonics, such as the elevation of Africa by 1200–1500 m during the Plio–Pleistocene, have been pieced together from the humble data of morphology (Smith, 1982 quoting King, 1962). But without numerical ages to refine and correct the calculated rates of movement, the issue of mechanisms, and thus of forces, has to remain in abeyance.

Chapter Three

ARCHAEOLOGICAL AND HISTORICAL SOURCES

*Johnson observed . . . that he had often wondered how it happened,
that small brooks . . . kept the same situation for ages,
notwithstanding earthquakes, by which even mountains have been
changed, and agriculture, which produces such a variation upon the
surface of the earth . . .*

Boswell, Life of Johnson

Samuel Johnson's observation arose during a discussion with Joshua
Reynolds, Edward Gibbon and others about a brook near Brindisi described
by Horace. It touches on the difficulties raised by archaeological evidence
and by historical sources other than eyewitness reports of earthquakes or
other ground movements. Even where artefacts are being used for dating
geological features their function as well as their age may raise problems
of interpretation: thus a wall now under water indicates coastal submergence
if it was a jetty but not if it was a breakwater. The problems are multiplied
when the evidence for earth movements is entirely a matter of inference, as
would be the suggestion that the breakwater was constructed because
subsidence had rendered a stretch of coast unacceptably subject to rough seas.

Some of the pitfalls of interpretation have been illustrated by Ambraseys
(1971) in connection with archaeological evidence of ancient earthquakes.
The evaluation of early earthquakes can err seriously if it is based on
experience with modern man-made structures or if it is assumed that the
survival of a few monuments indicates that the sites have remained free
of damaging earthquakes. What is more, earthquakes, unlike wars and
epidemics, appear to have little long-term social or economic impacts, at
least insofar as the Near East is concerned.

Ambraseys was making these points in response to the growing number
of papers and books in which earthquakes and volcanic eruptions are used

44

to explain gaps in archaeological sequences or to account for migrations. The prime example is the prevailing enthusiasm for the view that Plato's Atlantis represented Minoan Crete and its collapse was caused chiefly by the eruption of Thera and the attendant ashfalls and sea waves. At present this particular claim remains unsubstantiated, but such is the pressure of repetition that the hypothesis has come to be presented in many books and papers as fact. At the same time, workers sceptical of the volcanic story are exploring afresh the evidence for a seismic alternative (Downey and Tarling, 1984). Atlantis figures again in the Bimini dispute. In the 1960s, divers reported the presence of submerged limestone blocks in the Bahamas and claimed that they formed part of a Cyclopean roadway built by the inhabitants of ancient Atlantis. According to this theory Brimini Island is the surviving tip of the lost continent, a view first popularized by the mystic Edgar Cayce (1877-1945). There is every indication that the blocks consist of natural beachrock, and ^{14}C dating of the beachrock places its formation 2200-3500 years ago, about 7000 years later than Cayce suggested and at a time when sea level was $1 \cdot 9$-$2 \cdot 2$ m lower. The ensuing transgression coupled with natural settlement of the beachrock would readily explain its current depth (McKusick and Shinn, 1980).

ARCHAEOLOGICAL DATING

There is a clear overlap between archaeological and other sorts of dating when applied to geology. All the methods rely to some extent on stratigraphical principles so that, for example, a stratum will be considered to be younger than a sherd or radiocarbon-dated log beneath it. Again, the palaeontological approach, with its reliance on assemblages, lineages, extinctions etc. can be recognized in the procedures by which artefacts are employed to date and correlate cave, river and lake beds. Where archaeology scores, especially when combined with historical sources, is in the wide range of environments where it can be applied. Moreover, it sometimes has a much greater resolving power than other dating methods. A well controlled local pottery sequence may make it possible to date sherds to within a few decades. At the other extreme, the best one can do is to distinguish artefacts from the products of accidental fracture or wear and hence to separate material of human manufacture and dating from say the last 3 million years from items which might be older. Given the high cost of radiometric methods and the absence of suitable samples at many crucial sections, no indication of age, especially if it is gratis, is to be scorned.

Ruins whose original form is unambiguous on occasion supply useful limiting ages and also information on the geometry of deformation. Part of

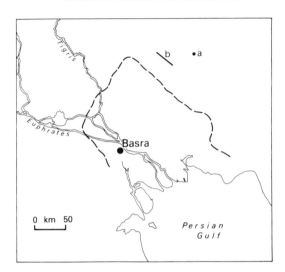

Figure 26 (a). *Lower Mesopotamia. The dashed line shows the position of the coast in 696 BC according to de Morgan (after Lees and Falcon, 1952). a indicates Dar-i-Khazineh, b the Shaur anticline.*

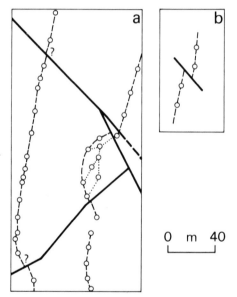

Figure 26 (b). *Fault offsets on ancient qanat (locally known as kyarizes) near Parou in the Kopet Dagh fault zone of Soviet western Asia. The full lines indicate faults; the qanat shown as dotted lines in (a) were dug to replace damaged qanat. The eastern part of the zone (b) is thought to be older than the central and western part (a), whence the greater displacement it displays. After Trifonov (1978) with permission from The Geological Society of America.*

the Great Wall of China built 400 years ago has been displaced by a fault with 1·45 m of right-lateral and 0·95 m of vertical slip (Deng *et al.*, 1984). At the Hisham Palace, built in the early eighth century near Jericho (Reches and Hoexter, 1981), two phrases of movement dating from the last 2000 years had been identified on the Jericho Fault, one of the two major strike-slip faults that bound the Dead Sea graben. The older event postdates a deposit which contains sherds dating from the early Iron Age (twelfth century BC) to the early Roman period (first century AD) but as all the sherds have been reworked only the youngest age is of significance. The second phase affects a deposit which contains sherds dating from late Byzantine to early Arabic times, that is the seventh to eighth centuries AD. Historical records suggest that the Hisham Palace was destroyed in AD 748 by an earthquake and it is reasonable to equate this with the fault displacement. Moreover, whereas the nature of the movements is not clearly indicated by the geological evidence, the present shape of the Palace suggests that they consisted mainly of left-lateral shear along a fault running roughly N 35° E. When the Palace was first surveyed in 1958 the irregularity of the rooms was blamed on its builders.

Equally valuable is the evidence of anticlinal uplift in Iraq provided by two canals dating from the Sassanian period (first and second century AD). Both canals were dug across the Shaur anticline, between Shush and Ahwaz in the Zagros foothills (Fig. 26). In 1955 it was found that one of the canals, which still carried water, had cut down some 4 m below its original bed (that is, a little over 2 mm/year), supposedly because its erosive power had kept pace with uplift. The other, which crossed a higher part of the anticline and whose middle section originally ran in a tunnel, was either too feeble in flow or too poorly maintained to survive, but at the crest of the anticline its bed had risen about 18 m in 4 km equivalent to 10 mm/year. As one might expect, these short-term averages are higher than the 7 mm/year of uplift calculated for coastal anticlines in the Zagros for the last 7000 years and the average of 1 mm/year proposed for the Zagros as a whole since the early Pliocene (Lees, 1955; Lees and Falcon, 1952; Falcon, 1974; Vita-Finzi, 1982).

The size of the archaeological sample will generally be greatly increased by excavation, and a true assemblage rather than one created by the accidents of redeposition may emerge if a living floor is encountered. At 'Ubeidiya, also in the Jordan Valley, river and lake beds now dip by as much as 75°. Excavation was prompted by the discovery of hominid fragments turned up by a bulldozer. The 'Ubeidiya formation is separated from later units by an angular unconformity (Fig. 27). The Formation has been subdivided, in order of decreasing age, into a lower lake series

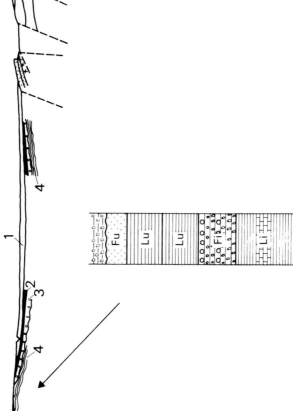

WSW

ENE

Figure 27. 'Ubeidiya Formation of the Central Jordan valley. Upper: Stratigraphic position. (1) Lisan Formation; (2) Yarmuk basalt; (3) Naharayim formation; (4) 'Ubeidiya Formation. Lower: Main constituent units. The Li and Lu units represent limnic (i.e. lake) deposits; Fu and Fi are fluviatile deposits. Based on Picard and Baida (1966), with permission from The Israel Academy of Sciences and Humanities.

(Li), a lower fluvial series (Fi), an upper lake series (Lu), and an upper fluvial series (Fu). The greatest number of artefacts—including a living floor—were recovered from the Fi deposits (for details, see Picard and Baida, 1966; Isaac, 1972; Horowitz, 1979). Beds 21–23 yielded choppers and flakes ascribed to Phase I of the Oldowan II. Beds 24–29, like the contemporaneous living floor, contained choppers and flakes, and also

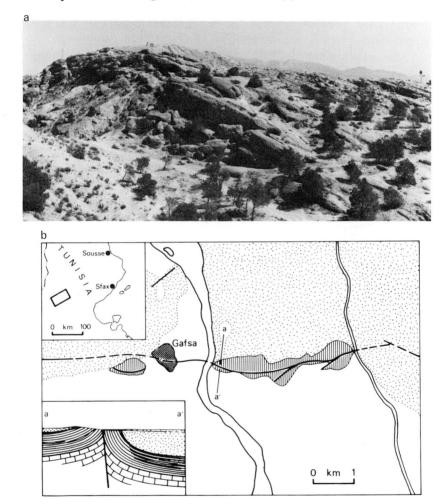

Figure 28. (a) *View of the folded gravels of Gafsa.* (b) *Geological setting. The stipple represents horizontally bedded alluvium containing Mousterian artefacts; the line-shading represents the folded gravels. Note the fault running through or near the gravels. The two deposits are also shown in the lower inset, a NNE-SSW section* (a-a) *based on a borehole data. After Castany (1955).*

spheroids, polyhedrons and picks, equated with Phase II of the Oldowan II which, to judge from K/Ar dates in the type localities in East Africa, spans the period between *c*. 1·5 and 2·5-3 million years ago. The artefacts in question occur about 70 m below the unconformity, which is followed by the Middle to Upper Pleistocene Naharayim and Lisan formations. Lavas younger than the 'Ubeidiya Formation have given K/Ar ages of 690 000 ± 140 000 years. In other words, the beds were tilted and dislocated at some stage between 1·5 million and 690 000 years ago. The result appears disappointingly imprecise until we recall that prior to the discovery of artefacts within it, the 'Ubeidiya Formation was ascribed by different authors to the Villafranchian (i.e. over 1·8 million or so years ago) and even to the Pliocene (say 7 million years ago). The lower limit provided by the basalts is perhaps even more informative, as no upheavals as severe as those responsible for the 'Ubeidiya structures appear to have occurred since the lavas were poured out.

Almost as dramatic and perhaps more contentious are the folded gravels of Gafsa, in Tunisia (Fig. 28) (Castany, 1955; Coque, 1962). In 1887 D. Collignon reported finding worked flints in the gravels of a small hill at Gafsa; by 1906 attention was being drawn to the steep dips displayed by the gravels and in 1911 a phase of folding was accepted as post-Acheulian in age. Detailed studies of the artefacts were not carried out until 1923,

a

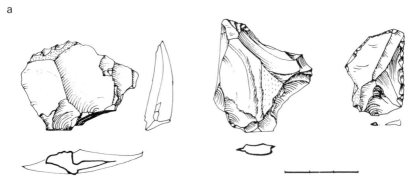

Figure 29. (a) *Artefacts from the folded gravels at Gafsa, Tunisia. Although the sample is too small for confident attribution to the Lower or Middle Palaeolithic, it confirms the presence of artefacts within the gravels.* (b) (opposite) *Artefacts from the tilted gravels near El Abadia (formerly Carnot) north of El Asnam, Algeria. 1-4 from the oldest unit, 5-8 from the intermediate unit and 9 from the youngest unit. None of the pieces is chronologically diagnostic and all of them could date from the Lower Palaeolithic although 9 is more probably of Middle Palaeolithic manufacture. Scale bars in centimetres. Drawn by Mrs L. Copeland.*

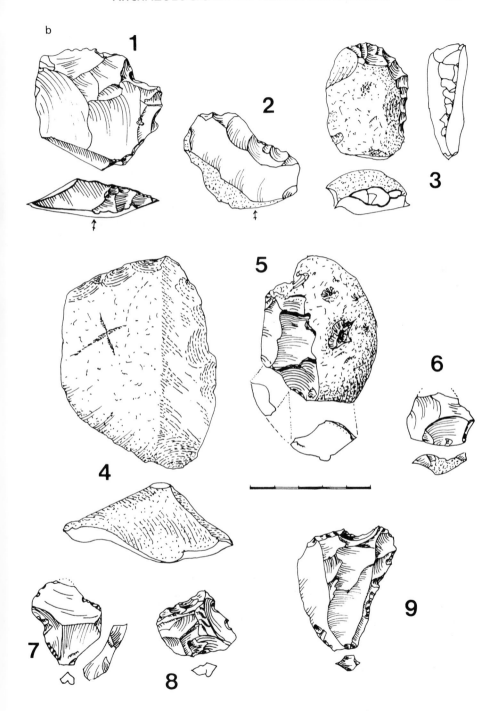

when Vaufrey showed that the material *in situ* was of Acheulo-Mousterian type, but in 1952 Gobert came to the conclusion that it was a late Acheulian industry.

There is only one reliable date for an Acheulean industry in Europe, namely $0 \cdot 429 \pm 0 \cdot 007$ million years for Torre in Pietra, in Italy, because volcanic rocks suitable for K/Ar dating are absent from the relevant sites. In Eurasia industries with a Mousterian 'aspect' start to appear about 100 000 years ago and had largely replaced the Acheulian by about 80 000-70 000 years ago. Although one observer was to suggest that the dips at Gafsa were purely depositional, the presence of the artefacts was not challenged until 1962 when it was suggested that they were not within the gravels but rather stuck onto the surface of the fold by a calcareous crust. The thought that reputable field workers should have made such an elementary mistake for three-quarters of a century is worrying, but a visit to the site provides reassurance: the gravels are rich in artefacts firmly embedded among the stratified pebbles.

Even the small sample illustrated [Fig. 29(a)] evokes a Lower rather than a Middle Palaeolithic context in the mind of an experienced prehistorian.* The fold is lapped by a valley fill which is underformed and which contains Mousterian artefacts and also a blade industry. The conclusion must be that folding occurred between about 1 million and 30 000 years ago, presumably in response to movement on a steeply dipping E-W fault. The estimate will be regarded as excessively vague by the outsider. To the geologist it is very informative.

The archaeological dating of tilted slope deposits at El Asnam, in Algeria, also produced a result which at first glance appeared unhelpful. Three generations of colluvium were identified on the southeastern flank of the anticlinal Pondeba ridge (King and Vita-Finzi, 1981; Thommeret *et al.*, 1983; Stein and King, 1984). The oldest deposit is nearly vertical, the second dips at about 45° and the dip of the youngest is roughly parallel to the channel floor. The two older units were found to contain Lower Palaeolithic artefacts with some Acheulean affinities; the youngest yielded material which, though by no means out of place in an Acheulian context, was more likely to come from a Middle Palaeolithic assemblage [Fig. 29(b)]. The simplest explanation for the archaeological and geological evidence is that progressive uplift of the anticline has led to progressive steepening of slope deposits on its flanks and began at some stage after the manufacture of the Lower Palaeolithic material $1 \cdot 0$-$0 \cdot 1$ million years ago. The previous best estimate would have had to be 'during the last 7 millions years' as this is the approximate age of the folded Late Tertiary rocks making up the Pondeba ridge.

In short, the onset of deformation at Gafsa and El Asnam can be put

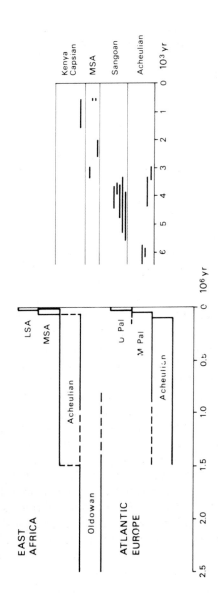

Figure 30. Left: Decreasing duration of successive gross cultural divisions in East Africa and Atlantic Europe. Adapted from Isaac (1972). MSA and LSA, Middle and Lower Stone Age; U Pal and M Pal: Upper and Middle Palaeolithic. Right: Time range of four cultural groups in sub-Saharan Africa. The lines represent ^{14}C dates for individual sites at three standard deviations. Note overlaps between and gaps within the groupings. From Coles and Higgs (1969).

at about 100 000 years ago. The scope for error is considerable and the apparent synchronism of the two folding episodes could well emerge as illusory. But the results should not be dismissed as worthless. Without the artefacts, both sets of deposits would have invited attribution to the Neogene, that is to say between 2 and 26 million years ago. Second, although the lower limit for the Acheulian is blurred, its upper limit is relatively crisp, and the Middle Palaeolithic industries that came to supplant it include features which are not difficult to recognize. Note that continuing deformation during and after Middle Palaeolithic occupation is not excluded by the field evidence. All that can be said is that any resulting tilt does not exceed the depositional dips that are nowadays regarded as usual in fluviatile deposits.

The resolution of archaeological methods will tend to improve as we approach the present, simply because the duration of gross stratigraphic divisions based on technology decreases (Fig. 30). Overlap between 'cultures' is of course a warning against the assumptions that there is a single, unilinear progression which will permit scattered events to be arranged in chronological order without serious difficulty. In general, the

Figure 31. Post-Roman terrace elevated to 8 m above the modern channel within the Pondeba gorge near El Asnam, Algeria.

geologist will gladly delegate the task of identifying archaeological material in the hope of thereby gaining an authoritative limiting date for the deformed deposits. Unfortunately many prehistorians are either too cautious to hazard a numerical age for small assemblages or inclined to use the relative terminology by which the Pleistocene was originally subdivided into four or five major glacials. In Europe this scheme begins with the Donau glacial and proceeds through the Günz, Mindel, Riss and Würm. It is still used by prehistorians and geologists alike even though the oceanic record suggests that the number of glacial episodes was much greater than five and the original terms were based on evidence from the Alps which is now being reassessed.

The process of uplift can be traced into historical times at El Asnam by reference to a deformed alluvial terrace whose regular gradient upstream and downstream of the ridge highlights the uparching it has undergone locally (Fig. 31). The terrace contains Roman sherds and overlies ruins from the baths of the Roman town of Tigava. Evidence for late incision in the headwaters of the Chelif had prompted the view that deposition ended about 1850, and ^{14}C dating of the fill in the Pondeba Gorge was to confirm this estimate: a charcoal sample at a depth of 60 cm had an age of 700 ± 80 years and another at 40 cm an age of 120 ± 70 years. Calibrated by reference to tree-ring tables, the upper sample is found to date from the period 1655–1950, so that the 5 m of uplift undergone by the terrace surface within the gorge and the resulting stream incision all took place in less than 330 years.

Figure 32. River terraces at Sadaich, Iran. Artefacts from the surface of the older (higher) fill are associated with a midden composed of marine shells dated by ^{14}C to 7300 years BP.

Prehistoric remains are of limited value in coastal studies, as they generally reveal little more than that a particular tract of land was dry during the period of occupation. This helps to explain why in Scandinavia the tendency has been to use shorelines for dating prehistoric sites rather than the other way round. The exceptions include the young *Littorina* shoreline of Finland, although it should be noted that the age of 5500 years obtained for it was derived from a Stone Age chronology built up largely from shoreline data (Zeuner, 1958; Siirinäinen, 1972). But things

Figure 33. Artefacts from the Sadaich site. Their early Neolithic character tallies with the ^{14}C age obtained for the midden. Scale in cm.

are very different if one can show where the site lay in relation to the contemporaneous shoreline. For example, shell middens, such as those now submerged on the coast of Nova Scotia (Grant, 1970), presumably lay close to the shore. At Sadaich, in southern Iran, some 16 km from the sea, shells from a midden composed almost entirely of *Meretrix* sp. and bordered by scatters of early Neolithic artefacts gave a ^{14}C age of 7300 ± 140 years (Figs 32 and 33). In view of ethnographic evidence to indicate that middens rarely lie more than 1 km from the collecting grounds, the contemporaneous shoreline lay close to 10 m above its present position, a value that, after allowing for 10 m of eustatic sea-level change, is in fair agreement with other evidence for Holocene tectonic uplift thereabouts at about 3 mm/year (Vita-Finzi, 1981).

Some supposed fossil shorelines are probably middens rather than beach assemblages, an error all too easily made when one is dealing with the remains of a single meal (rather than repeated occupation) and if no artefacts or charcoal are associated with the shells. One thereby loses the geological benefits of a shoreline deposit but in compensation gains a shell sample of uniform age which was collected live and will therefore give a ^{14}C age which refers to the corresponding sea level.

An unusual example of coastal occupation which yields clear evidence of shifts in the shoreline is to be found in Saudi Arabia, where Al 'Ubaid dwellings made of reeds are encrusted with barnacles (Fig. 34) and thus show that they formerly stood in the intertidal zone. In one example (Bibby, 1970) net emergence amounts to 4 m since 500 BC. But coastal sites of the classical age are the most familiar example of archaeological items used to identify deformed tracts of coast, and the Mediterranean is the sea that has produced the largest number (Richter, 1958, p. 56). The extent of submergence is clearly a minimum measure of the relative rise in sea level as the buildings originally stood an unknown amount above high water. At Carthage a roadway built on the shore in the first to third centuries AD is now $0 \cdot 25 – 0 \cdot 40$ m below water level, which suggests (by analogy with modern building practice) that submergence amounts to about $1 \cdot 25 – 1 \cdot 4$ m (Yorke and Little, 1975). Flemming (1969) had previously concluded that any eustatic change in the western Mediterranean during the last 2000 years could not have exceeded $\pm 0 \cdot 5$ m, the accuracy within which his site surveys were effected.

As Flemming observes, 'the functional nature' of submerged remains has to be taken into account if one is to obtain reliable results. Fish tanks, slipways and moles were built with foundations at least partly below sea level; quarries were probably not cut below sea level. Careful field methods combined with corrections for waves, the tide and atmospheric conditions help to ensure dependable results. Fishponds at the Roman/Byzantine site

Figure 34. Reed impressions in clay at an uplifted Al 'Ubaid (fifth millennium BC) site on the Arabian shore of the Persian Gulf.

of Lambousa, on the north coast of Cyprus, can thus be used to show that the sea has retreated 1 m since they were in use (Dreghorn, 1981). The same principles apply to later historical times, witness the mooring rings set up on the coast of the Maritime Provinces of Canada between 1717 and 1737 and now 37 cm below the highest tide (Grant, 1970).

The scope for ambiguity remains. A striking instance is the archaeological evidence for recent movements on the Israeli coast (Neev *et al.*, 1973; Horowitz, 1979). According to one groups of workers, there are clear indications of submergence to depths of up to 40 m less than 3800 years ago (and probably nearer 700 years ago) followed by uplift of the coast along an offshore fault (Fig. 35). Shell beds dominated by *Glycymeris violacescens* are found at various sites along the coast at heights ranging

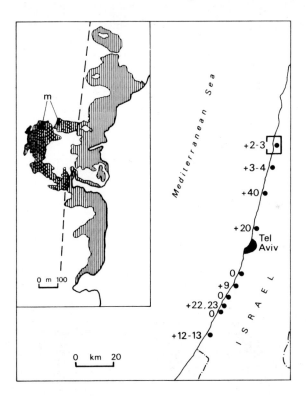

Figure 35. Israel coast. Location of dated shoreline sites and their heights in metres. The northernmost is Caesarea. Inset: *features around the Herodian harbour in Caesarea.* Hatching: *rocky bottom;* stipple: *artificial rocky bar* interpreted as partly collapsed Herodian breakwater; m: *masonry with lead-cast joints;* broken line: *inferred fault line. After Neev et al. (1973) with permission.*
© *1973 Macmillan Journals Ltd.*

from 3 to 40 m. The shells have given ^{14}C ages of between 2120 ± 200 and 3800 ± 210 years; some of them overlie Roman and Crusader sites and are interbedded with sherds of which the youngest are 700 years old. Similar shells found on the modern shore give ^{14}C ages of 1150–1520 years. If this is accepted as a measure of the time taken for the dead shells to be deposited on the beach, the elevated beds are 600–1650 years old. Submergence of the offshore zone is indicated by the presence of lead joints in the breakwater of the Herodian harbour of Caesarea, built about 10 BC, now over 10 m below sea level. A line detected on air photographs bounds rocky patches on the sea floor and is taken to mark the fault east of which uplift occurred.

Other workers dispute this interpretation of the field evidence and explain the shell beds by fluctuations of sea level. A compromise solution is to ascribe initial submergence to a post-Roman transgression, with the ensuing uplift a truly tectonic effect. And there are those who maintain that most of the coastal sites dating from the last 4000 years indicate less than 1 m of sea level displacement. Significant tectonic activity was detected near the Haifa-Qishon graben and, although Caesarea showed differential submergence, this could be explained by local slumping or faulting. In

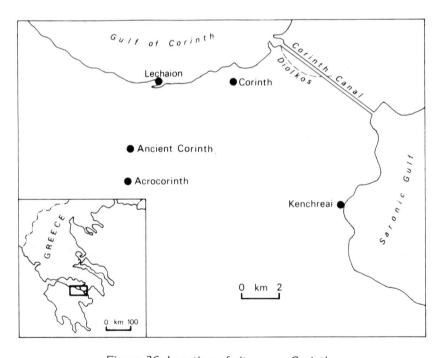

Figure 36. Location of sites near Corinth.

Figure 37. Beachrock overlying the northern end of the diolkos or ramp that was used for crossing the Isthmus of Corinth in antiquity.

short there is no indication of a continuous coastal fault. The elevated shell beds are presumably to be dismissed as middens.

Whatever its outcome, the dispute over this coast serves to remind us that vertical movements are not necessarily unidirectional. The northern end of the *diolkos* or ramp by which Greek ships crossed the Isthmus of Corinth before the famous canal was excavated (Fig. 36) bears the signs of submergence followed by emergence. The blocks making up the slipway, which probably dates from the fifth century BC, are covered by sandy beachrock (Fig. 37). To judge from its composition this represents intertidal or shallow-water conditions whereas the top of the bed now lies 50 cm above high water. Once again the submergence can be ascribed either to subsidence or to a continuation of the postglacial trend. The ensuing emergence, however, cannot be explained otherwise than by uplift (Vita-Finzi and King, 1985). The pillars at Pozzuoli likewise indicate at least two phases of movement.

CULTURAL CLUES

At one extreme the data yielded by archaeology are considered sufficiently dependable for statistical assessment and the construction of trend surfaces.

Figure 38. Roman mausolea at Mselleten, Libya. Note rotated capstones.

At the other they remain tantalisingly inconclusive. The 'staircase fault' at Qumran, south of Jericho, has been cited as a product of historical seismicity and even ascribed to the earthquake in 31 BC, but sceptics suggest it merely represents local subsidence. There is also a body of information whose potential has still to be realized. The Roman mausolea at Mselleten in Libya (Fig. 38) both display a rotation of the capstone which is doubtless seismic, but all that can be said at present is that the event or events are post-Roman. As Oldham pointed out with regard to similar effects during the 1897 Indian earthquake, actual twisting is possible only on very soft ground where conditions approach those of a liquid, and the most probable explanation is that waves arrived from

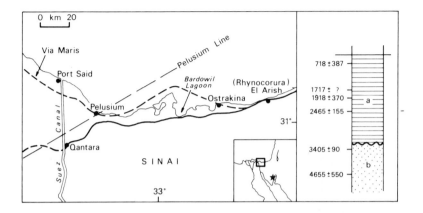

Figure 39. Recent tectonic activity postulated for Sinai. Left: *NW coast. The Via Maris was in use* c. *2700-500 years ago following uplift of the coastal belt and before it was interrupted by subsidence along the Pelusium and Bardawil troughs.* Right: *Profile of the upper part of the sedimentary record in Solar Pond (star on small map) southwest of Eilat, showing* (a) *lagoonal facies overlying* (b) *open-sea facies. After Neev and Friedman (1978) with permission.* © *1978 The American Association for the Advancement of Science.*

many directions while the structures were being shaken (Richter, 1958, p. 56).

Where the events themselves are a matter for surmise the conclusions are bound to inspire little confidence. But they may provide a useful test of existing views or, at the very least, point to areas that might reward further study.

One of the structural models proposed for the recent evolution of the southeastern Mediterranean identifies a Sinai sub-plate which is bounded site of Pelusion, is seen as a north-northeast-trending compressional zone about 60 km off the southeastern Mediterranean coast which runs onshore site of Pelusion, is seen as a north-northeast-trending compressional zone about 60 km off the southeastern Mediterranean coast and runs onshore near the Bardawil Lagoon northeast of Cairo (Fig. 39). Recent activity on the line 3000-2700 years ago has been inferred mainly from archaeological sources. The oldest traces of human activity on a straight lagoon bar is Persian pottery of about 2700 BC, and this was taken to indicate uplift of the bar above sea level before that date. Likewise zones of subsidence are identified from submerged sites such as Ostrakina, which was occupied from the Roman period to Mamluk times, say 500 years ago. In addition, the main coastal route from Egypt to Syria and Mesopotamia was replaced by one further north between Persian and

Mamluk times, but this was later abandoned supposedly in response to coastal subsidence.

It is evident that the climate of opinion at the time of writing has to be considered when reviewing such evidence. The Pelusium line conforms with a plate-tectonic model of the region. Anticlinal uplift of the Shaur canals could be postulated independently of regional models, even if the uplift rates are valuable for testing competing models. On the other hand, the geosynclinal picture of the Persian Gulf and its Mesopotamian extension that was in vogue in the 1940s and 1950s implied subsidence of this trough in compensation of the uplift displayed by the Zagros, and fluvial geomorphology at the turn of the century postulated stream incision in response to a fall in base level or uplift of the land. Two items of archaeological evidence appeared to confirm both contentions: submerged irrigation works at the head of the Gulf, and, inland, alluvial deposition followed by channel trenching by as much as 5 m during the last 6000 years (Lees and Falcon, 1952; Macfadyen and Vita-Finzi, 1978).

Much of the coastal evidence can be explained by the postglacial rise in sea level coupled with the compaction often exhibited by delta deposits. Detailed analysis of borehole data showed that an estuary extended at least 250 km above the present head of the Gulf in recent times and that much of lower Mesopotamia is floored by river alluvium, which amounts to reinstating a view first put forward by W. K. Loftus in 1855 to the effect that delta growth had pushed the coast towards the southeast since late prehistoric times (Fig. 26). Marine molluscs in the borehole sample had prompted the idea of repeated incursions by the sea due to phases of subsidence but their co-existence with freshwater species is more consistent with estuarine conditions. No crustal movements are required to harmonize this narrative with the field data.

The fluviatile record is also explicable without appeals to uplift or subsidence. The entire region has been affected by successive phases of river aggradation and downcutting during the Quaternary. Slight shifts in the seasonal incidence of rainfall provide a more economical explanation for the sequence than would earth movements in such a tectonically complex area. The fate of the Neolithic village of Beidha, near Petra in Jordan, appears to be an exception. The site receives less than 200 mm of rainfall a year, the nearest perennial spring is 5 km away and the nearby wadi contains no soil. The location would be much improved if the alluvium that formerly occupied the wadi were still present, as Beidha would thereby acquire cultivable land and also water from wells tapping the groundwater of the valley floor (Raikes, 1967). Given the unstable nature of the Jordan Rift, a tectonic explanation for the damaging episode of erosion is plausible. The present author objected to it because the movements were entirely *ad hoc* and lacked independent confirmation, whereas the climatic mechanism was at least regionally valid. A visit to the site in March 1985 has prompted a recantation. There is no need for any explanation at all. Beidha is surrounded by land which produces a fair barley crop by rain-fed, technically primitive agriculture even today.

No environmental alternative has yet been advanced to account for agrarian decline since pre-Hispanic times in a part of coastal Peru where, according to several archaeologists, a major irrigation canal was rendered useless by progressive tilting of the region (Browman, 1983). Unlike the two crossing the Shaur anticline, the La Cumbre Canal was central to the region's prosperity. It carried water from the Chicama valley to the Moche valley over a distance of 70 km, but now displays reversed gradients in at least seven places. According to exponents of the tectonic hypothesis, the Chimu had 2000 years of experience in canal engineering before they built the La Cumbre Canal in the eleventh to the thirteenth centuries AD. The effects of regional tilting and uplift were not confined to artificial irrigation: rivers deepened their beds and the water table fell, so that the canal intakes had to be shifted and in due course abandoned. Survey work on two structures north of the Moche Valley support the idea of tilt, but the evidence of the canal itself is threatened by two other interpretations: that it was built to create work and that it was badly designed from the outset.

The most persuasive application of a tectonic hypothesis to archaeology has been in the Indus Valley. The Harappa culture with its offshoots occupied a large area in northwestern India and West Pakistan in the third and second millennia BC. Its decline, or at any rate the end of its principal cities, was for many years attributed to invasion, although some authors saw in climatic desiccation a factor that was at the least contributory. Raikes (1964, 1965) then suggested that coastal uplift affected the natural drainage system and, besides disrupting river and coastal communications, led to ponding of the Indus waters and, in consequence, extensive silt deposition. The silt gradually engulfed Mohenjo-daro and other sites in the adjoining parts of the Indus floodplain.

Lyell (1837) has described the extensive ponding that in 1819 followed uplift of Allah Bund, in the northern Rann of Kutch (Fig. 40).* The Makran coast of Pakistan, like its Iranian counterpart, has been subject to spasmodic uplift throughout the Quaternary, and the abandonment of the Harappan ports of Sutkagen-dor and Sotka-koh has been attributed to the effects of this uplift coupled with sediment accumulation.

---

*During the Kutch earthquake there was uplift north of the Allah Bund and subsidence south of it over a distance of 90 miles (145 km). The net difference in elevation was about 10 m (Oldham, 1926).

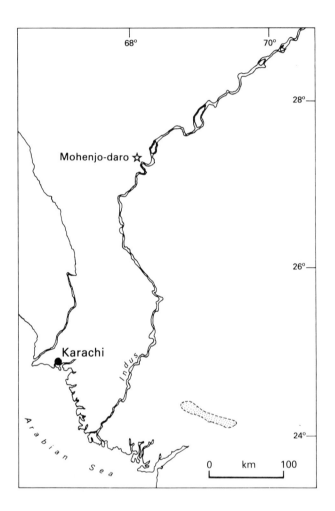

Figure 40. The Allah Bund (stipple) and Mohenjo-daro.

The results of drilling into the valley floor at Mohenjo-daro also support the uplift hypothesis by revealing, to a depth of 15 m below the floodplain, a series of silty clays of the sort to be expected where deposition occurs in standing water rather than under flood conditions.

Although the theory has been well received, the field data on which it rests have been challenged by one experienced observer (Lambrick, 1967). In his view, much of the material attributed to deposition in ponded Indus waters consists of disintegrated mudbrick and wind-blown silt. Mohenjo-daro was finally abandoned when the Indus changed its course in a manner which starved of water the lands around the city.

These are matters to be resolved by careful excavation and through comparative study of contemporaneous sites occupying different settings. The lack of any feature unambiguously identifiable as the dam responsible for silt deposition is the main obstacle to acceptance of the Indus record as evidence of crustal deformation; which amounts to saying that, in the absence of field evidence, the hypothesis remains little more than an intriguing possibility.

Chapter Four

MAPS AND MEMORIES

The island tilts. The tourists go mad.
E. M. Forster, The Longest Journey

The sources brought together in this chapter bear directly on ground deformation by way of eyewitness accounts and the comparison of successive surveys or instrumental readings. But the directness does not always bestow significance on the results: unlike geological and archaeological data, such measurements and observations are confined to the last three millennia—in some areas to the last few decades—and the rates and patterns of deformation they reveal may be unrepresentative of the long-term or regional trends. What is more, the relationship of the observed movements to the local structure can prove elusive, so that the benefits of a precise record of events are not fully realized.

It would seem to follow that the information is most useful when appended to the narrative of geology and archaeology. But there have been occasions when direct observation of ground movement have contradicted the accepted map of deformation by revealing past activity in areas now quiescent or the converse. In addition, several examples of active deformation have emerged from levelling intended solely for map revision. At a more analytical level, however, surveys are being set up to investigate known or suspected movements, a trend promoted by growing interest in prediction for scientific and practical purposes and also by the need to limit the extent of repeated surveys in view of their high cost. Direct monitoring of movement by permanent devices is also increasing in importance, especially now that field methods can be supplemented or replaced by remote sensing from satellites.

EYEWITNESS REPORTS

The inclusion of eyewitness accounts in a chapter dealing with geodetic techniques may seem odd, not least because instrumental measurement can give extremely high precision and its reproducibility can be specified. But the human observer should not be scorned even where he has apparently been superseded by optical and electronic systems: he can be critically selective and consequently a very effective source, especially in recording ground deformation during earthquakes, and he is mobile and cheap to run.

The Persian seismic record goes back to the Assyrian period in the thirteenth century BC when an earthquake which damaged Nineveh was the subject of correspondence So far it has yielded a handful of references to geological effects. In the fourth century BC, for example, a Greek author held that Rhagae (Ray) was thus called because the root of the very 'rent' is *rhag* and the ground was commonly rent by earthquakes thereabouts (Ambraseys and Melville, 1982). In AD 856 an earthquake in the Alburz was accompanied by extensive ground deformation in the mountains, and some of the streams in the area were dammed by landslides. In 1641 an earthquake near Tabriz set off extensive rockfalls and landslides, and fissures opened. In 1838 an earthquake in Sistan was accompanied by liquefaction of lowlying areas which rendered caravan routes unsafe for years afterwards. Such reports are given far less prominence than accounts of loss of life and damage to dwellings and monuments, and they are of limited value even when trustworthy. A common problem is that one cannot readily distinguish between tectonic faulting and the fissuring that often accompanies landslides. Not that modern accounts are invariably superior: the difference between primary and secondary faulting remains an obstacle to the assessment of earthquake effects, and many seismologists still lack field experience (Ambraseys and Melville, 1982, p. 171, n. 21).

Historical sources benefit greatly from being combined with skilled mapping from air photographs and on the ground even if centuries have elapsed since the earthquake struck. This applies to the fault break associated with a major earthquake which occurred east of Birjand (Iran) in 1493, although the lineament probably owes nothing to earlier displacements on an existing fault. The Bozqush reverse fault, which can be traced for 2 km, is a Quaternary fault which was reactivated in historical times (Berberian, 1976) and notably in 1879.

The value of such historical reports, once their reliability has been tested, lies not only in the activity but also in the gaps and pauses they reveal. Taken in conjunction with other symptoms of seismic activity, the reports may make it possible to estimate the intensity of the event and the probable

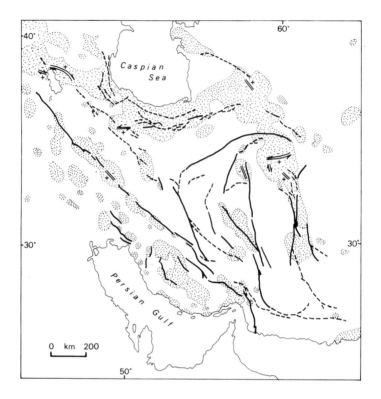

Figure 41. Parts of Iran affected by destructive earthquakes since AD 700 (stipple) and faulting during this period (double lines with indication of uplifted and downthrown sides or sense of transcurrent motion). Major Quaternary faults (bold lines) are also shown. After Ambraseys and Melville (1982).

location of the epicentre. The length of the fault break is of course a valuable clue to magnitude. It turns out that none of the faulting associated with historical and other recent earthquakes in Persia occurred on a major Quaternary or Tertiary fault (Fig. 41). Almost all the recorded instances were connected with relatively minor fault zones of recent age located largely in northern and eastern Iran. The Zagros, often regarded as highly seismic, shows no major or great earthquakes: activity consists largely of many shallow, relatively small earthquakes with little evidence of large-scale faulting. (One of the few exceptions, in 1909 at Silakhur, was associated with faulting over a distance of over 45 km and with vertical displacements of 1–2 m.)

The seismicity itself is of interest. Apparent gaps emerge both for the historical period as a whole and for the twentieth century. In other words a seismic map based on recent events is misleading. The deficiencies of the record will not explain the gaps as they occur in populated as well as desert regions and, conversely, large events at any rate are recorded even for remote areas. The implications for neotectonics are clear. To use an old adage, absence of evidence is not evidence of absence.

Iran is notoriously seismic. Arabia is commonly thought to have a very low seismicity. The earthquake that struck Yemen in December 1982 appeared to be an exception, especially as it gave rise to extensional fissures up to 10 km long. But a fairly continuous record of medium magnitude shocks is indicated over the past 1200 years by Arab histories, local chronicles, the writings of passing travellers and official archives (Ambraseys and Melville, 1983). The evidence has gone some way towards correcting the view that any recent activity on the Arabian peninsula has been confined to its Red Sea coast.

Eyewitness reports do more than extend the geological record: they facilitate its interpretation by bringing to light processes which accompany ground deformation and can then be used as clues to earlier events which are imperfectly recorded in the field evidence. The first isoseismal map was probably the one drawn by Sir William Hamilton (Guest, 1982) for the Calabria earthquake of 1783. The classic account of the 1897 Indian earthquake by Oldham (1899) included many perceptive comments including the suggestion that, whereas some deep-seated fissures had propagated upwards from below, others had propagated downwards from the surface. Oldham also described seismically activated sand vents from which sand and water were ejected (cf. the 'sand volcanoes' on p. 173). He also inferred deformation from interruptions in surface drainage unaccompanied by faulting and from reported changes in the 'aspect' of the landscape, including improved intervisibility between two heliograph stations. In his report on the 1926 Kutch earthquake Oldham also remarked on ground slopes which were not immediately noticeable but were later revealed by changes in the character of the surface deposits (Oldham, 1926).

On the other hand, Oldham was vague about the driving mechanism, and it took another perceptive field worker, H. F. Reid, to make the crucial observations. The fundamental association between faulting and earthquakes was not recognized in the nineteenth century. Lyell described only two instances of seismic faulting in the *Principles of Geology*; Oldham ascribed the various effects of the 1897 Indian earthquake to a subterranean disturbance or 'bathyseism'. John Milne suspected that fault displacement was the source of elastic waves and, following the Mino-Owari earthquake of 1891, B. Kotō concluded that faulting was not an effect but the actual cause of the earthquake Richter, 1958. Kotō's report was published in 1893, but it was not until the work of Reid on the 1906 California

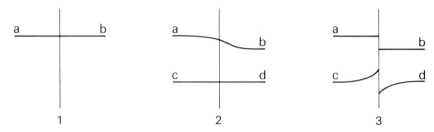

Figure 42. The elastic rebound model of H. F. Reid, formulated after the 1906 California earthquake (1) *unstrained conditions, straight feature* ab *(e.g. a road) crosses the fault.* (2) *horizontal strain deforms* ab. *A new straight feature* cd *is set up across the fault.* (3) *a major earthquake relieves the strain by strike-slip movement and elastic rebound. The two parts of* ab *are now offset;* cd *shows curvature as well as offset.*

earthquake that this view gained general acceptance (Boore, 1977). For Reid reported displacements of up to 6 m for several hundred kilometres along the San Andreas Fault. He was also able to compare geodetic surveys run before and after the earthquake. Finally, and perhaps most important, he provided a model — the elastic rebound theory (Fig. 42) — which showed how the field data could be explained and hence what parts of it deserved detailed field study.

The 1964 Alaska earthquake provided ample scope for exploratory assessment of movement combined with purposeful search for diagnostic items (Plafker, 1965, 1967). The deformation that came to light was of great interest to students of mountain and continental evolution. It occurred along a belt 800 km long which bordered the Aleutian Arc, one of the great volcanic areas and deep ocean trenches that surround the Pacific, and, what is more, it occurred above sea level (Fig. 43).

The pattern that emerged can be summarized as a seaward belt of uplift, with a maximum of 10 m on land and 15 m offshore, and, to the northwest, a belt of subsidence averaging 1 m and with a measured maximum of 2·3 m. On the coast the evidence for movement included over 800 measurements of the displacement undergone by barnacles, mussels, algae and other organisms whose ecology was well understood. *Balanus balanoides*, the acorn barnacle, proved especially valuable because it leaves a clear mark on the shore with a sharp upper boundary and the results could be checked against the upper limit of new barnacle colonies elsewhere on the Pacific coast. (It turned out that this is close to annual mean high water.) Some of the effects of the 1899 Yakutat Bay earthquake had been assesed by comparing the height of dead and living barnacles; useful confirmation of the 1964 results was obtained from the establishment of a new barnacle line by the

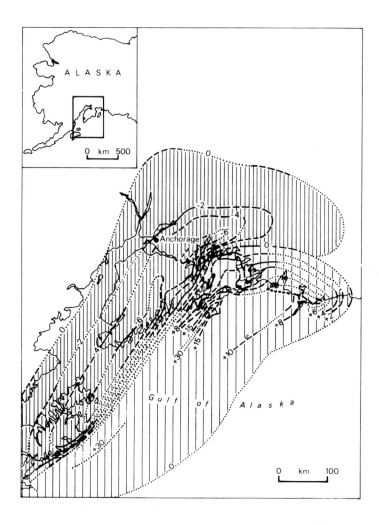

Figure 43. Regional deformation that accompanied the Alaska earthquake of 27 March 1964. Measurements in feet. Heavy shading = subsidence, light shading = uplift. After Plafker (1967).

end of the field season. Other evidence included topographic features, such as new beach terraces, lagoons and raised sea caves and drained lagoons, and the displacement of roads and jetties.

The extensive warping indicated by the observed displacements prompted a rigorous search for surface faults. The shoreline data combined with air reconnaissance in due course revealed two reverse faults in the southwestern

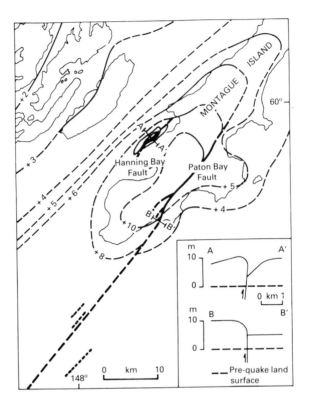

Figure 44. Uplift (in metres) and faulting on Montagu Island during the 1964 Alaska earthquake. After Plafker (1965).

part of Montague Island (Fig. 44). The precise age of the features remained in doubt as they were found four days after the earthquake and away from the shore, where the barnacle line supplied direct evidence of displacement. The faults had to be inferred from the topographic breaks, shear zones, fissures and fallen trees. Submarine scarps indicated their probable extension offshore, and seismic sea waves suggested that considerable movement had occurred on the sea floor.

For much of their length the two faults follow lineaments which are clearly visible on air photographs, but the age of undisturbed trees along the fault lines suggests that there has been no major displacement for at least 150–300 years. The faults show a maximum dip-slip component of some 8 m but they are not regarded as the primary features along which the earthquake occurred (Plafker, 1965, 1967). Unlike most major tectonic boundaries, the fault lines mark no significant change in rock lithology.

Figure 45. Graben near the crest of the anticline over the reverse fault at El Asnam, Algeria.

The observed movement is inadequate to explain the regional uplift, especially in the seaward belt. The focal region defined by the aftershocks is too extensive. Finally, the epicentre of the main shock was nearly 150 km away from the fault traces and not at a location which could lie on a down-dip projection of the fault planes. The absence of a major fault separating the major zones of uplift and subsidence thus shifted attention to other hypotheses to explain the pattern of deformation and seismicity (see Chapter 8). It is worth noting that the existence of such a fault had been consistent with the position of the epicentre and with a preliminary fault-plane solution.

With the exception of simple linear features which were demonstrably once continuous, straight or level, such as fences and roads, field evidence of ground deformation however accessible it might be has by no means become a matter of routine inspection. Many features are open to different interpretations even at the descriptive level. The distinction between faults and landslips is often troublesome. Secondary faults produced, as Oldham suggested, by surface deformation may indicate tensional conditions when

Figure 46. Ephemeral evidence of thrust faulting at El Asnam, Algeria.

the primary mechanism is compressional. Immediately after the 1980 earthquake at El Asnam, tensional features were far more prominent than the direct effects of reverse faulting (Fig. 45), which, apart from telescoped irrigation pipes, consisted of frail thrust scarplets likely to be destroyed by the first rains (Figs 46 and 47). It is conceivable that an earthquake in the area in 1954 produced thrust features and that these were overlooked whereas normal faults were observed and mapped (King and Vita-Finzi,

Figure 47. Telescoped irrigation channel parallel to the direction of shortening at El Asnam, Algeria.

Figure 48. A matter of interpretation.

1981). Again, irrigation lines crossing the main scarp appeared to indicate sinistral strike-slip movement, but this proved to be an illusion produced by ground disturbance close to the fault. Likewise, the normal fault breaks observed at Thessaloniki after the 1978 earthquake could have been mistaken for reverse faults (King *et al.*, 1981).

Even without the problems of field interpretation already mentioned, the notorious unreliability of most eyewitnesses (Fig. 48), especially after any kind of natural calamity, is sufficient warning. Anyone who has tried to assess the effects of an earthquake by interviewing the local population will know that agreement over what has happened may be difficult to obtain even between a fisherman and his wife (Vita-Finzi and King, 1985). Confusion will be increased if there is any reversal in the displacement after the main event.

It is of course rare for eyewitness reports to bear the entire burden of a chronological study of deformation. In reconstructing the events of the 1857 earthquake of central and southern California, historical records were combined with tree-ring analysis, which made it possible to identify trees that had been disturbed by faulting (Sieh, 1978b). Some segments were found to have moved by 6-9·5 m, others by only 3-4 m. Rupture was detected along at least 360 km of the fault, but much of it has since remained dormant and only the stretch northwest of Cholame has been affected by moderate earthquakes associated with fault slip. This kind of information is of value for the analysis of recurrence intervals and for evaluating seismic moment.

Provided it is used with circumspection, myth and tradition may also have something to contribute. In New Zealand, historical earthquakes have left their marks inland and on the coast. In 1835, for example, the west Wairarapa Fault tilted the Rimutaka Range, and its fault scarp ranged in height from 2·7 m to almost nothing (Stevens, 1974). At Turakirae Head five raised beaches can now be distinguished. The youngest was uplifted 2·5 m in 1855. The next one represents 5·8 m of uplift. It has been dated to about 1460, as Maori tradition set down in the early 1900s recorded that some 18 generations earlier a great earthquake known as Hao-whenna (or land swallower) was accompanied by coastal uplift at Miramar Peninsula nearby. Higher beaches can be dated only by comparison with beaches elsewhere for which there are ^{14}C ages. The younger two represent uplift at about 10 m/1000 years (Fig. 49).

Somewhat more explicit is the report that Swampy Cree Indians in Arctic Canada, who know that emergence is occurring, argue over whether it is the product of uplift or a fall in sea level. On the other hand, the one item that could lead to an uplift rate—a place name which means a river junction at the water's edge and which refers to a feature now 3 m above sea level—is of limited value as it is attributed to 'old people' dead for an unspecified period (Andrews, 1970, p. 8).

Human observers have proved lazy and unreliable, especially over features whose absence is of significance. As Ambraseys (1978, p. 189) remarks, faulting in the Near and Middle East is not as rare as it was once thought to be, presumably only because it has tended to go unobserved. Yet even where the burden of reporting the past is shouldered mainly by the geologist and geomorphologist, the eyewitness retains one key function: that of reducing the ambiguity inherent in many seismic records (see below, p. 112) by careful observation of the deformation that accompanied the earthquake.

MAPS AND INSTRUMENTAL SURVEYS

The comparison of individual features recorded on maps or photographs remains a useful device especially in the exploratory stages of a study. The New Zealand tectonic narrative is particularly well illustrated by

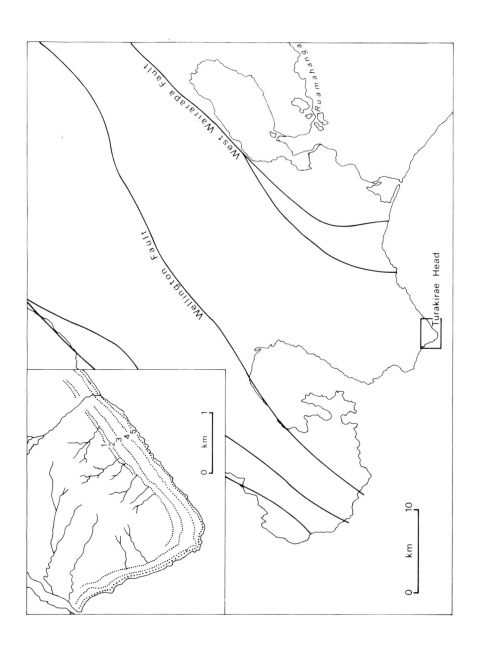

a

b

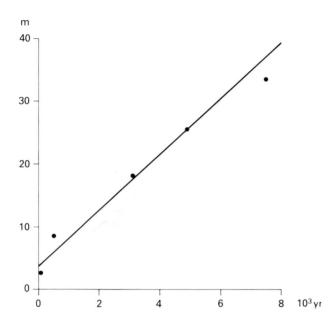

Figure 49. (a) (opposite) *Faults that have moved in the last 15 000 years in the southwest of South Island, New Zealand. Inset shows fossil beaches on Turakirae head at about 25, 22, 16, 8 and 2 m above sea level. After Stevens (1974).* (b) *Plot of net uplift against age at Turakirae Head. The terrace heights given by Stevens (1974) have been corrected for the eustatic effect using global sea-level data in Flint (1970).*

photographs, some of which cannot be bettered by instrumental surveys because the features in question are now covered by roads and buildings or have been eroded away. In Algeria, two successive editions of the 1:100 000 topographic map were used to test the proposition that the El Asnam earthquake had been accompanied by warping of the valley floor. On the 1934 sheet there was, of course, no trace of the lake that formed in 1980 (Fig. 50), but the 1890 edition showed a marsh occupying approximately the same position as the lake. This was presumably the remnant of an earlier such warping feature. Although the finding cannot be validated, or fully exploited, except by drilling to uncover and date any organic lake beds, the cartographic comparison was important by virtue of the minimum recurrence interval it indicated — namely over 90 years — and the impetus it gave to the search for ground evidence of seismic folding

Figure 50. Lake brought into existence by warping across R. Chelif during El Asnam earthquake of 1980. Note uparched terrace deposit bordering the lake and compare its elevation here with that in Fig. 31.

in Algeria and elsewhere using geological as well as historical data (King and Vita-Finzi, 1981; Stein and King, 1984).

It is very unusual for the immediate products of crustal movements to be revealed by the simple comparison of successive topographic sheets if only because such discrepancies tend to be blamed on the inadequacies of the earlier survey. But geodetic techniques, which amount to the map data without the map, are a major weapon in the hunt for recent deformation, and the growing number of high-precision surveys set up solely to investigate earth movements have still to be anchored to survey stations originally set up for the construction and revision of topographic maps.

The results of instrumental measurement are unusually valuable in areas where other sources are either too insensitive or impracticable. In Australia, for example, low rates of movement give rise to slight topographic features which are soon obscured, and there are few, well dated late Cainozoic deposits or surfaces which can be used as reference levels (Wellman, 1981). Where Holocene deformation is well documented, however, measurements of current movement will help to show how far long-term average rates remain valid and, indeed, whether there has been any change in the type of deformation. In central Honshu, Japan, geodetic measurements are in agreement with data bearing on the last 2 million years. For instance, the Akaishi Mountains bear every sign of Quaternary elevation, and the surveys show they are now rising at 3 mm/year. Nevertheless, changes in tempo over the period of measurement can be detected, notably an increase in the tilt rate of the Suruga coast since 1973, which raised the possibility that readjustment to the 1944 Tonankai earthquake was superimposed on the long-term pattern (Thatcher and Matsuda, 1981).

If the interval between observations is sufficiently long, and movement sufficiently rapid, relatively inaccurate data can still serve a useful purpose. In New Zealand this proved to be the case when a survey made between 1878 and 1884 came to be compared with geodetic surveys made between 1938 and 1971 in order to assess deformation on the Wairau Fault. The early survey had a mean trianglar misclosure of about 5″ of arc. Few angles were common to it and the later surveys. Comparison was thus based on shear components of the surface strain because this method does not depend on scale or orientation. Deformation was taken to be proceeding at a uniform rate. To illustrate the results, the region crossing the Wairau Fault gave a principal strain of $0 \cdot 35 \pm 0 \cdot 09$ microradians/year, with the azimuth of the most comprehensive principal axis at $90 \pm 7°$ E. Comparison with the results obtained within areas on either side of the fault showed no discontinuity at the fault and thus supported the evidence of microearthquakes

Table 2
Errors in determining recent crustal movements (mm/Δt)

Line length (km)	Horizontal movements		Vertical movements			
	1979	1990	1979		1990	
			a	b	a	b
0·1	0·3	0·03	0·15		0·1	
1·0	0·7	0·1	0·45		0·3	
10	15	2·0	2·0	55	1·5	3·5
100	50	7·0	7·0	170	5·5	13
1000	100	20	75		15	
9000	300	100	200		75	

Simplified after Prilepin (1981). The columns represent the time (Δt) between measurements required to give a particular precision. For example, in 1979 horizontal movements over a length of 1 km expressed in mm/year would have required resurvey after an interval of 0·7 year. In the last four columns, a = geometrical levelling and b = trigonometric levelling. For the 1000–9000-km range, very long-base radiointerferometry is employed.

for current quiescence on the fault. Thus, besides confirming broadly the geological evidence for a principal horizontal strain axis at 110–115° E for the last 20 000 years, the exercise lent support to the view that strain was accumulating as elastic energy (Bibby, 1981).

Recent crustal movements operate on scales that range from the regional to the very local. Various attempts have been made to subdivide the spectrum. One scheme recognizes four levels, starting with movements in active faults, near volcanoes etc. (distances of about 0·1–5 m), moving on to tectonically active regions (several to a few hundred kilometres) and large tectonic blocks of the lithosphere (hundreds to thousands of kilometres) to culminate in global deformation (Prilepin, 1981). Table 2 shows the time between successive measurements required to attain a particular error level at different linear scales. The table compares the state of affairs in 1979 with expected improvements by 1990.

The main improvements are likely to continue to be in the measurement of distance. Angular measurement between trigonometric stations has had an accuracy of 2–3 p.p.m. (parts per million) since 1850; distance measurement improved dramatically after 1950 when the tellurometer was introduced, and is now 1–5 p.p.m. (Wellman, 1981). The major problem faced by trigonometric levelling, that of refraction, is being tackled by the development of refractometres; in Table 2 the corresponding error for 1990 is assumed to be 0·1″ (seconds of arc). The problem lies not so much in the future, when progress is assured, but in the exploitation of existing data. In drawing attention to the deficiencies of levelling during the last century, Bendefy (1968) notes that one such survey in Hungary showed a remarkable resemblance to the map of the 'beginning of the blossom time of the small-leaf lime-tree', because the results of the survey were strongly influenced by microclimatic features.

The presentation of results is more commonly graphical. The use of isolines to link points of equal displacement, though risking spurious smoothing, permits regional patterns to be grasped. As in the study of displaced classical sites, the construction of polynomial surfaces is valuable especially when the data are scattered or irregularly distributed provided the underlying method is borne in mind. Flemming (1971) found no simple correlation between the age and the depth of his archaeological sites in the Aegean, but a third degree surface which he fitted to the rate of displacement of the sites produced low residuals and helped him to identify regions affected by similar vertical movements. Third-order surfaces also made it possible to construct a map of contemporary vertical movements in Canada using 5046 relevelled segments, 39 pairs of lake-level readings and data from 47 tide gauges (Vanicek and Nagy, 1981).

The geodetic investigation of ground deformation immediately after an earthquake, though often logistically problematic, offers the advantage that the area to be surveyed can be identified without much difficulty. According to Oldham (1899), the triangulation executed after the 1897 Indian earthquake was the second such survey: the first had been done in Sumatra in 1892. The maximum vertical displacement reported by Oldham was 12 feet (*c.* 3·7 m); this was away from the area where movement had been at its greatest, but the changes there could not be established because all the stations were disturbed.

When the southern Afar was affected by an earthquake swarm in November 1978, a topographic survey revealed at least 1·6 m of extension since 1972-3, when the network was set up, and levelling showed that parts of the inner floor of the rift had subsided over 70 cm since 1972 (Abdallah *et al.*, 1979). The first set of geodetic measurements had used nine of the 22 original stations. Further measurement revealed that most of the movement was confined to the inner 10 km and that this was bordered by narrow bands of contraction. The vertical movements had accentuated the existing topography. Subsidence affected a zone 4 km long and 2 km wide. The bands subject to contraction had been uplifted by as much as 20 cm. There was also some indication that the faulting was superimposed on general uplift with a wavelength of about 20 km.

The 1983 earthquake at Coalinga, in California, had a magnitude of 6·5 M_s. Although one of the aftershocks, 15 km west of the main shock, was accompanied by surface faulting along a distance of 5 km, the main event was not. As at El Asnam, a lake came into existence just upstream of an anticlinal ridge which follows the strike of the fault plane revealed by the hypocentre and aftershock data. The river-bed profile of two rivers which cross the ridge was plotted from maps based on surveys conducted in 1955-6, and compared with the form of river terraces that border them as observed in the field or indicated on the same maps. The pattern that emerged from the comparison (Fig. 51) supported the hypothesis that repeated activity on the fault had led to terrace uplift of about 10 m in 2550 years.

The effects of the 1983 event could be assessed by comparing a profile run after the earthquake with benchmarks surveyed in 1972. Comparison with the 1937 edition of the relevant sheet also proved helpful: it showed that subsequent water extraction by pumping could not account for the observed deformation. The modern earthquake had produced uplift to a maximum of 0·5 m. In the absence of deformation between earthquakes and the predominances of events similar to that of 1983, the total uplift indicated by the terraces requires a recurrence interval of 125 years. But the calculation is complicated by differences in the path followed across the anticline by the geodetic and terrace profiles. Once they are drawn to the same horizontal scale, one can see uplift in parts of the terrace profile

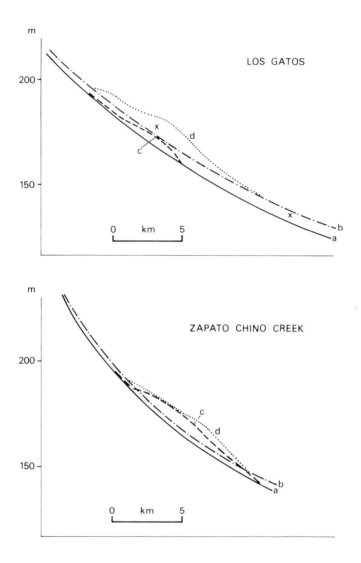

Figure 51. Profiles along two streams transverse to the strike of the fault
responsible for the 1983 Coalinga earthquake: (a) undistorted bed profile, (b)
undistorted flood terrace level, (c) distorted river bed profile and (d) distorted
river terrace. Based on field measurement and map data. The crosses mark sites
dated by [14]C. Note warping of river beds and terraces in both streams. After
King and Stein (1983).

where the 1983 event produced subsidence. Stein and King (1984) conclude that more than one fault has been responsible for deforming the terraces and, after correcting for the effects of this 'broad uplift', put the uplift of the 1983 event at 5·5 m and thus the recurrence interval at 350 years.

Upstream and downstream of the anticlinal ridge the Los Gatos Creek has a meandering course. Where it crosses the ridge the river is relatively straight, and this pattern was taken to tally with the map of uplift. It would clearly be of help in the exploratory stages of a survey if one could use stream geometry as a dependable guide to recent increases or decreases in channel gradient.

A recent study of the response of alluvial rivers to deformation in Louisiana and Mississippi shows that the hope is not unrealistic (Burnett and Schumm, 1963). As the channels of alluvial rivers are floored and bordered by alluvium, they are relatively free to adjust to changing gradients (and other disturbances). The study began in two areas shown by repeated geodetic levelling to be undergoing uplift and composed of deformed Tertiary strata. Pleistocene and Holocene alluvial terraces, alluvial valley floors and projected channel profiles (or plots of channel bed height against valley length rather than channel length) showed convexities where they crossed the uplifts. In addition, the present streams display changes in morphology which also reflect uplift, but the picture is complicated by variations in channel size. For instance, of the streams crossing the Wiggins uplift, the Pearl River, which is large, has a relatively straight longitudinal profile, though downstream of the axis of uplift it has cut down to leave its former floodplain as a low terrace. The Bogue Homo Creek, which is smaller, displays an anastomosing and relatively straight channel upstream of the axis of uplift, presumably in response to a reduced gradient. Downstream the channel has cut down below the former floodplain; it is also more sinuous than upstream, and braided reaches have formed where incision has led to increases in sediment load (Fig. 52).

The deformation observed at El Asnam and Coalinga showed that subsurface fault movement could lead to surface folding and hence that geodetic data can shed light on concealed structures. Resurvey also has potential for the analysis of volcanic activity, especially when allied to the monitoring techniques discussed in Chapter 5. The pioneering work was done by R. M. Wilson on Kilauea, Hawaii, in 1935. Repeated levelling showed that successive eruptions were accompanied by marked uplift and subsidence of the volcano. For example, in 1924, there developed a depression 10 km across and up to 4 m deep. Between 1967 and July 1968, following inflation of the volcano over several months, there were a series of summit eruptions. Nevertheless, the association between activity and deformation is not simple. The summit eruptions of 1967–8 were not

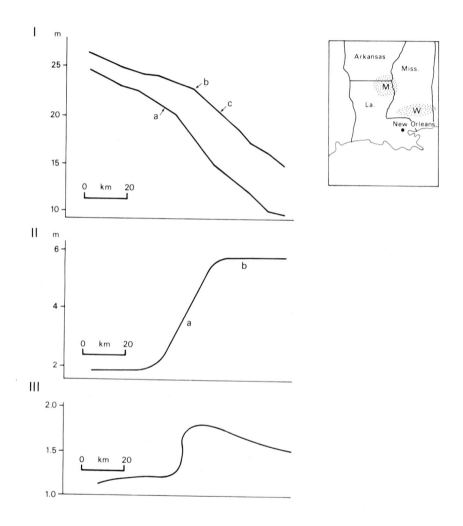

Figure 52. Effect of the Monroe uplift (M in inset) on the Big Colewa Creek.
I: Channel profile (a) and valley profile (c). Note increased gradient and depth
of the channel downstream (right) of the uplift axis (b). II: Channel depth, with
the middle reaches showing active entrenchment. The lower reaches are already
entrenched but the upper reaches are not. III: Sinuosity at a peak in the middle
reaches. W on the inset is the Wiggins uplift, where similar effects are to be
seen. After Burnett and Schumm (1983) with permission. © 1983 The American
Association for the Advancement of Science.

Figure 53. Levelling on Mount Etna to evaluate topographic changes. Photograph by courtesy of Dr J. E. Guest.

accompanied by significant surface displacements, whereas three later periods of flank eruption—in August and October 1968 and February 1969—coincided with subsidence and contraction of the summit region. Such observations, together with the shifting geometry of deformation when it does occur, suggest that the magma reservoir on Kilauea is not a single chamber but a complex network of chambers. In other words, the geodetic evidence is likely to contribute more to prediction by what it reveals about the inner workings of the volcano at issue than by any simple premonitory pattern of deformation (Decker and Kinoshita, 1972; Dvorak *et al.*, 1983).

The plumbing of Mount Etna was indeed the primary concern of a study of ground movements associated with volcanic activity over a period of 4 years (Murray and Guest, 1982). An 11-km traverse of benchmarks was set up in July 1975 and reoccupied twice a year to record any vertical displacements. The traverse was supplemented by 32 tilt stations each consisting of three benchmarks 30–60 m apart in a nearly equilateral triangle. (They are referred to as 'dry' tilt station to distinguish them from those dependent on a device containing a fluid to provide an artificial horizon.) The work, arduous and tedious enough (Fig. 53), was hampered throughout by the activity of vandals who damaged or destroyed survey markers.

Little vertical movement was observed except on fresh lava flows, which consistently showed downward movement attributed to compaction and sliding. As the subsidence exceeded 1 m in an area at least 2 km across in the space of 14 months, fears that it could represent incipient caldera collapse proved unfounded. There was no simple pattern of inflation before eruption and deflation after eruption. Ground deformation served to indicate the location rather than the timing of potential eruptions. For example, the Vulcarolo–Torre del Filosofo region south of the summit underwent inflation after September 1976, and a new crater opened in April 1978 about 600 m north of the centre of inflation. The probable explanation is that there is no large-scale storage of magma in the upper few kilometres of the volcano, and magma rises at a variable rate from a depth of about 20 km. But fracturing of the walls of the central conduit or conduits can happen 2 years before the eruption or even longer. The petrological data support this scheme.

Ground deformation understandably enough has received less attention on non-basaltic volcanoes characterized by infrequent, explosive eruptions. Mount St Helens is one of the exceptions (Lipman and Mullineaux, 1981; Chadwick *et al.*, 1983). Systematic measurement, which began early in April 1980, was prompted by fracturing and bulging of the summit, and was performed using aerial photographs, levelling by theodolite combined with electronic distance measurement, and gravity measurements.

By mid-April the rate of bulging locally attained 2·5 m/day, and the high point of the bulge had risen 150 m above its original position. On Kilauea any extension or uplift before eruptions amounts to a few millimetres per day. The bulging could be explained by intrusion of a body of dacite magma, and emplacement of a new lava dome seemed a likely outcome. In the event the oversteepened north flank failed on 18 May when an earthquake of magnitude 5 triggered off a landslide and a set of explosions which removed 2·7 km³ of material from the volcano. The reasonable conclusion must be that slope failure in its turn precipitated the eruption by suddenly removing the confining pressure that had held it at bay. Geodetic surveys emerge as all the more important where the internal workings of the volcano thus interact with gravitational forces.

The search for movements which will give advance warning of earthquakes is also being conducted vigorously in many parts of the world (see Chapter 9). The observations are geared increasingly to a specific explanatory hypothesis, as where dilatancy is seen as a crucial component of the seismic process. In others there is merely a belief that a change in the tempo or character of deformation might indicate the onset of an earthquake (Tsubokawa *et al.*, 1964; Castle *et al.*, 1975). Areas classified as seismic gaps may thus attract more attention from surveyors than their seismic neighbours. For example, the Shumagin Islands of the eastern Aleutians is one such gap for which a great earthquake is expected to occur within the next 20 years, and evidence of ground tilt is being monitored by survey lines already relevelled nine times between 1972 and 1983 and with sight lengths of 30 m or less (Beavan *et al.*, 1983).

The most familiar instance of current deformation is probably the San Andreas Fault. Creep on the fault came to light at a winery near Hollister; it had totalled 10 cm between 1948 and 1956. In 1957 the Coast and Geodetic Survey set up monuments near the winery and, once the original rate had been confirmed, the study was extended to two triangulation networks, one near Hollister and the other 75 km to the southeast (Whitten and Claire, 1960). This was apparently the first time geodetic measurements had been employed for recording crustal movements. Displacement on the fault averaged $0 \cdot 2 \pm 0 \cdot 01$ m for the former (1930–51), and $0 \cdot 05 \pm 0 \cdot 01$ m for the latter (1932–51). The range of values in each case was considerable, with a few stations registering motion on the fault contrary to the majority trend. In short, the study revealed the merits and limitations of geodetic work; it also demonstrated the variation to be found in displacement rates along individual faults.

In 1964 G. P. L. Walker suggested, in the light of field data, that the fissures supplying Iceland's volcanoes totalled a thickness of about 400 km, equivalent to an annual expansion rate of $0 \cdot 5$–$1 \cdot 0$ cm/year for the past 50 million years. Repeated precision levelling since 1966 has permitted vertical movements to be followed on some of the major active faults in the island's rift zones. Despite climatic obstacles—levelling is confined largely to June, July and August—consistent results point to subsidence of the central part of the rift zone relative to the bordering areas of between $0 \cdot 4$ and $1 \cdot 0$ cm/year according to location. Horizontal displacements are as usual more elusive but significant extensions were observed in the eastern rift zone as early as 1967–70, when 6–7 cm of widening occurred (Decker *et al.*, 1971; Gerke, 1974; Tryggvason, 1974). Less conclusive were the results obtained on the International Karakoram Expedition in 1981, which had as one of its aims to find the monuments set up in 1800 for the Survey of India and to determine any displacements in their original position. On the assumption that India is currently being driven northward relatively to Eurasia at a rate of about 50 mm/year, linear displacements of 4 m were anticipated. In the event, the errors in the original survey were found to exceed any discrepancies between it and the 1981 survey (Bilham and Simpson, 1984).

The supposition that the Alps are still undergoing uplift has thus been bolstered by the results of repeated precision levelling at an interval of 50 years. Analysis of the data for the Basel-Chiasso line showed symmetrical doming of the Swiss Alps at the crest at rates of $0 \cdot 44$–$0 \cdot 76$ mm/year, the highest value being for St Gotthard (Schaer *et al.*, 1981). Uplift rates were also obtained by fission-track dating of apatites from the local granite-gneiss rocks, which permitted deformation of the 120°C palaeoisotherm to be traced. At that temperature the apatite fission-track 'clock' is restarted; rocks whose fission-track memory had been wiped clean by the high temperatures of metamorphism would thus record the data of their uplift above the 120°C isotherm. The results for the last 7 million years are $0 \cdot 38$–$0 \cdot 54$ mm/year.

The Fennoscandian region is already provided with long-term estimates of uplift rate which depend on a variety of Quaternary and Holocene sources. In this case it is hoped that departures from the average in the very recent record will be detected, as this would shed helpful light on such matters as the extent of 'relaxation' (see below, p. 148). For example, in Finland uplift isobases have been constructed using precise levelling data from 1892–1910 and after 1835 (Kiviniemi, 1981). The maximum

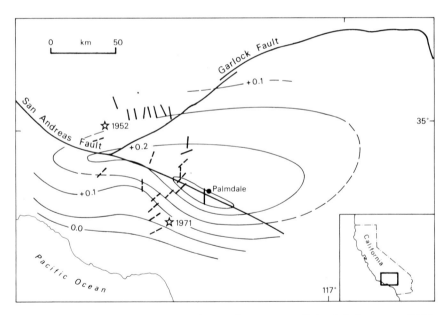

Figure 54. Uplift (in mm) in the Mojave Desert near Palmdale between 1959 and 1974. After Castle et al. (1976) with permission. The bold lines show the orientation of the compressive strain axes determined from triangulation data for 1952-63 and 1959-67. The broken lines have been corrected by subtracting the normal secular shear strains. The stars show epicentres of earthquakes in 1952 (M = 7·7) and 1971 (M = 6·4). After Thatcher (1976) with permission. © 1976. The American Association for the Advancement of Science.

uplift is 9 mm on the Gulf of Bothnia. The results agree with the amount of coastal advance indicated by maps of the mid-seventeenth century.

Without exaggerating the differences, one can distinguish between such purposeful surveys and those that are carried out on a more speculative basis. The Adirondack Mountain massif prompted such a survey because, being larger and higher than other 'uplifts' in the Interior Lowlands, it gave rise to the suspicion that uplift might be occurring now (Isachsen, 1975). Comparison of levelling records obtained in 1955 and 1973 showed that uparching has indeed occurred, with uplift of 40 mm at the centre and subsidence of 50 mm at the northern end giving a total of 90 mm in 18 years.

A more celebrated instance of unexpected, localized uplift is the Palmdale bulge of California (Fig. 54). In 1976 it was found that 12 000 km² of southern California had been undergoing uplift since 1959 with a maximum total of 0·25 m at Palmdale. The zone in question extended east of the junction between the Garlock and San Andreas faults in an area that had experienced little seismicity since 1932. The levelling in question met first-order standards and was linked to a tide gauge. In an attempt to identify the mechanism responsible for uplift, the orientation and average rate of shear straining during the time period at issue were calculated from repeated triangulation surveys. Most of the crucial strain apparently occurred between 1959 and 1963. At other times the pattern conforms to the picture to be expected from aseismic slippage on the San Andreas fault at 30-40 mm/year north and south of the crucial area. Plotted on the uplift map, the compressive strain axes for 1952-63 and 1959-67 are seen to orient themselves perpendicular to the uplift contours (Castle *et al.*, 1976; Thatcher, 1976; Wyss, 1977). One possible conclusion is that, if anything, the uplift has locked the fault further and thus impeded stress release by slippage. At all events the horizontal and vertical deformation was attributed to strain accumulation across northward-dipping thrusts like the San Fernando Fault.

By 1977 the bulge had collapsed but the uplift area was found to extend further southeast than originally thought and a second uplift peak of 45 mm had become manifest. In 1980 the entire phenomenon was dismissed as the spurious effect of systematic error in the calibration of the levelling rods.

As in California, the discovery of recent movements from geodetic data would hardly come as a surprise in Israel. The novelty lies in the pattern thus revealed. Comparison of first-order levelling data for 1962 and 1969 in Galilee and between 1959 and 1966 in the Negev gave differences in height which ranged from −65 to +55 mm for the 7-year period. Where averaging the results would have suggested stability, plotting them on geological cross sections and maps showed that many known structures were still active. In Galilee, for instance, numerous changes of sign in the observed displacement across faults suggested that some of the faults were still active, and the general pattern indicated relative sinking of the structural backbone with respect to the coast and the Jordan Rift. Two traverses across the Negev also showed a close link between structure and displacements, and again pointed to an inverse relationship between topography and recent movement. Gravitational or other effects appeared not to be implicated in this unexpected result (Karcz and Kafri, 1973, 1975).

On the coastal plain of Israel, first-order levelling for 1962 and 1969 gave differences of up to 158 mm, equivalent to about 20 mm/year. Plotted

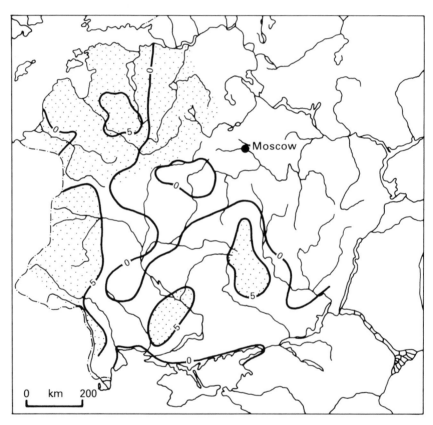

Figure 55. Rate of recent vertical movements (in mm/year) in the western USSR based on discrepancies between levelling data obtained in 1883 and 1912 and renewed levelling in 1953-9. The stippled zone is undergoing subsidence. The survey lines totalled 35 477 km and were tied to sea-level records at eight stations. After Mattskova (1967), with permission from Keter Publishing House, Jerusalem Ltd.

on the polygons defined by the levelling routes, the results hint at a block structure related to the E–W normal faults revealed by gravity surveys. As in Galilee and the Negev, there is a tendency for structural lows to rise and highs to fall. Besides their significance for the geophysics of Israel, these findings are in accord with widespread evidence for continental block structures which has been revealed by geodetic studies in other areas and especially in Japan (Bilham and Beavan, 1979). The blocks are 5–50 km long. As block boundaries are weaker than continuous continental blocks, strains are of course likely to be concentrated there.

It remains to be seen whether the vast number of geodetic changes already detected ultimately yield patterns of geological or geophysical value. For instance, it has been claimed that there is no area of consequence between the Baltic and the Black Sea in which secular movements are absent, and that a distinction can be drawn between two kinds of movement (Fig. 55). The first affects major crustal blocks (such as the Byelorussian–Lithuanian and Ukrainian basement outcrops), characterized by uplift for almost the whole of geological history, and the Dnieper–Donets and Black Sea depressions, with a tendency to subside. The second is observed along major, meridional wave-like folds of the crust which can give rise to fairly abrupt transitions between adjoining zones of uplift and of subsidence. A wave-like pattern (with a wavelength of either 250–300 km or 80 km) has also been proposed for recent movements along part of the coast of Japan (Mescheryakov, 1967; Zuchiewicz, 1984).

Chapter Five

MONITORING CHANGE

Geodynamics . . . has been a highly speculative subject for about a hundred years and it is not likely that this situation will change during the next hundred . . . partly due to the fact that the time intervals in which "something happens" are of the order of millions of years . . .

A. E. Scheidegger, 1963

The instrumental methods discussed in this chapter are used for tracing crustal movements which are currently taking place. They thus differ from geodetic techniques only in degree: even where continuous readings are obtained, attention will tend to focus on net change over a specific period— if only in order to see how far long-term trends have persisted—so that the claim to continuous monitoring is somewhat spurious. In any case, where does the present day pass into history? Yet the distinction between inference and direct observation deserves to be made. By analogy with medicine, geological methods reveal the pattern of growth, of fracturing and of superficial scarring; archaeology and geodesy permit a (partial) case history to be built up; but the need remains for direct consultation in order to keep track of subtle but cumulative changes and with increasing reliance on instruments to measure items ('shallow', 'regular') which would formerly have been assessed in the subjective light of experience.

STRAINMETERS AND TILTMETERS

Faults and folds suspected of vigorous activity are tempting subjects for direct gauging. A strainmeter can supplement the average rates of slip supplied by geology and geodesy with a continuous record of change. A tiltmeter should show whether the dips displayed by a fold, basin or dome are changing and, if so, whether deformation is continuous.

The solid Earth, like the oceans, is deformed periodically by tide-generating forces. Near the coast the resulting bodily tides are complicated by variations in the load on the ocean floor and in the local gravity field produced by the oceanic tides. The strain is in the order of 10^{-8} ($1\,\mu$m/100 m) and its measurement thus requires instruments with a resolution of 1 in 10^{10}. For a full assessment of the deformation, several strainmeters oriented in different directions are required.

The Benioff gauge, devised in the late 1930s, employed a long quartz rod which was mounted horizontally and fixed at one end. Temperature, humidity and pressure effects limited its sensitivity to changes in strain of 1 in 10^9, and the distance monitored was confined to the length (20–30 m) of the cylinder. Similar devices used metal tubes and wires. The mid-1960s saw the introduction of laser interferometers (Fig. 56) covering lengths of up to 1 km and capable of detecting microseisms with an amplitude of 1 in 10^{10} to i in 10^{12}. Earth tides are evidently well within the scope of such devices and, if necessary, can be weeded out from the record (Vali and Bostrom, 1968; Torge, 1990; Melchior, 1978).

But the results are by no means unambiguous. Besides the problems created by background and instrumental disturbances, there are oscillations whose nature remains unclear. Without necessarily subscribing to the view that vertical crustal movements are all oscillatory, one can see why some geophysicists feel that 20–30 years of observation are required before genuine secular movements can be identified. There is of course an alternative: to understand and therefore to make theoretical allowance for tidal and kindred effects. But it is not yet a practicable course of action.

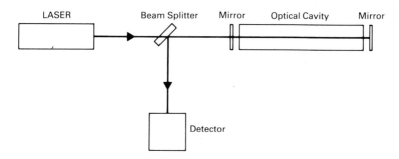

Figure 56. Principle of the laser strain meter. Interference patterns are produced when light reflected from the mirror at the centre meets light from the mirror at the right in the detector. When deformation changes the distance between the mirrors, the pattern is modified.

Progress is likely to be slow because most strainmeters are intentionally sited well away from potential sources of tectonic disturbance. Yet this has on occasion proved an advantage. In 1974 a wire stainmeter which had been operative since July 1971 at an underground field station at Nelson, New Zealand, recorded an abrupt change in strain amounting to $1 \cdot 2 \times 10^{-5}$. In the absence of any local seismicity the event was ascribed to 20 cm of creep on the Waimea fault, which is 5 km away at the surface but could be closer at depth. Some evidence of dilatancy about 20 months earlier was also obtained from the strainmeter trace (Gerard, 1975). But such instances of serendipity are few. Real gains can come only from applying dedicated instruments, even if lower in resolution, to specific structures or zones (Fig. 57).

A wide range of devices have been employed to this end. For example, a differential transformer was used between 7 October 1965 and 9 January 1966 to evaluate slippage on the Hayward Fault where it crosses the Memorial Stadium at Berkeley, California. G. D. Louderback had suggested in 1942 that, as the material bordering the Hayward Fault was more likely to deform plastically than as an elastic solid, deformation

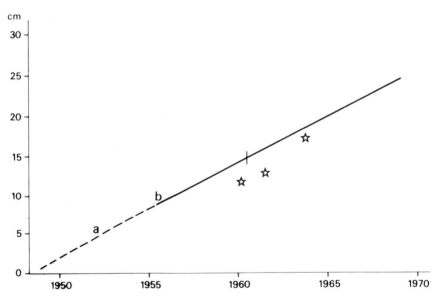

Figure 57. Cumulative creep on the San Andreas Fault near Hollister. Measurements start in April 1956 (b); *broken line* (a) *is estimated and originates at completion of the winery. 1949-60 after Steinbrugge* et al. *(1960); beyond 1960 after Meade (1971). Stars represent earthquakes of M $\geqslant$ 5.*

would be by slow creep; yet prior to the 1965-6 measurements fault slip or creep had been observed at only three other locations in California and perhaps nowhere else. The average rate of creep at Berkeley turned out to be about 5 mm/year. Being continuous, the record also permitted comparison of fluctuations in the speed of movement with the incidence of earthquakes during the corresponding period. The lack of any significant correlation between slippage and seismicity was an important piece of evidence in support of the view that the slippage represented the accumulation of strain energy in the brittle zone at depth rather than a response to seismic activity near the surface (Bolt and Marion, 1966).

Simple wire-type creepmeters have amassed practically continuous records of slip spanning 12 years on the San Andreas Fault. Seasonal and other variations emerged, but comparison with local rainfall gave no obvious pattern, although at some sites both the onset and the end of rains appeared to have some effect on creep rates (Schulz *et al.*, 1983). Moreover, different instruments do not always give the same results. Some non-wire creepmeters fail to record small events or do so only if sited close to the disturbed zone (Johnson *et al.*, 1976; Burford, 1977; G. A. M. King, 1983). Likewise tiltmeters with a resolution of 10^{-8} radians (i.e. $\sim 6 \times 10^{-7}$ degrees of slope) may be affected by rainfall. In the English Lake District, measurements lasting 9 months revealed statistically significant correlations between tilt and variations in the local water table (Edge *et al.*, 1981). Thus a change in tilt of 1 microradian (1 μrad or 10^{-6} rad) corresponded with a change in groundwater level of about 1 m. What is more, some of the changes in response to heavy rainfall displayed a rapid onset lasting a day followed by a decay lasting 10 days or so. In a seismic area such a problem could well be mistaken for precursory tilt.

It is of course the search for such warning signs of imminent earthquakes that motivates the vigorous research into tilt recording in Japan and China as well as the USA. Although the underlying mechanism is still disputed, the fact remains that changes in the direction of ground tilt have been reported a few days before several earthquakes of magnitude 3 or greater on the San Andreas Fault, and tilt is one of the phenomena being monitored closely in China with prediction in mind. How far the effects account for premonitory unrest or panic among animals has still to be determined.

The case for a continuous record of tilt on active volcanoes is if anything better, because the connection between tilt events and eruptions has been easier to demonstrate (UNESCO, 1972). On Mount St Helens electronic tiltmeters as well as various surveying techniques have been used with success in forecasting nine eruptions between December 1980 and the end of 1982. For several weeks before the eruptions the lava dome was seen to expand, on one occasion by as much as 32 m, presumably because viscous magma had risen in the feeder conduit. The rate of deformation increased before the eruptions to the point that eruptions were successfully predicted 3-19 days in advance (Chadwick *et al.*, 1983). On Etna, a small eruption in January-March 1974 was preceded by 3 weeks of fluctuations in the tilt of an area 3 km south of the central crater. By using an increase in the radial component of tilt as a token of inflation in the summit region, the geometrical changes obtained by repeated distance measurement could be used to test alternative models for the magma chambers in the volcano and thus to improve the quality of later forecasts (Wadge *et al.*, 1975).

The tiltmeter, which consisted of a spirit level mounted in a tube monitored for position on two orthogonal axes, was placed in a borehole 3 km south of the central crater. Three weeks before the eruption the

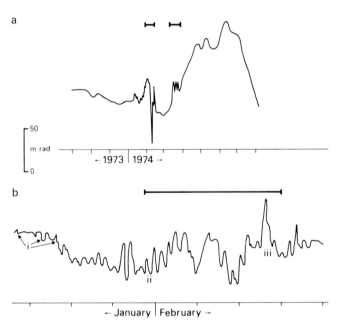

Figure 58. Mount Etna: tiltmeter record for 28 September 1973 to 8 August 1974. (a) Smoothed trace (sampled at weekly intervals) for tilt tangential to the central crater. Abscissa subdivided into months; positive change signifies ENE up. (b) Non-smoothed trace of radial component just before and during the first phase of the 1974 eruption. Abscissa subdivided into 5-day intervals. (i) Series of small deflations corresponding to the start of seismic activity felt at the volcano. (ii) Large amplitude tilt fluctuations (rapid inflation followed by deflation) which continue until after the start of eruptive activity. (iii) End of the first phase of activity. Bars in (a) and (b) denote phases of the eruption. After Wadge et al. *(1975) with permission. © 1975 Macmillan Journals Ltd.*

summit inflated and a series of fluctuations in tilt began [Fig. 58(b,i)] at about the time that seismic activity was recorded at the volcano. They may represent fracture propagation from the magma reservoir to the eruption site. The ensuing large-amplitude fluctuations [Fig. 58(b,ii)] may represent pulses of magma moving into the flanks of the volcano. They occurred before and a little after the start of eruptive activity. A series of large changes in tilt [Fig. 58(b,iii)] marked a brief renewal of lava effusion. A period of calm was followed by relatively minor tilting and renewed eruptive activity. Its close was followed by rapid summit inflation and then a reversal in tilt [Fig. 58(a)]. The last remains puzzling; the rest of the record at the very least yields some insights into magma inputs and withdrawals.

On Kilauea, Hawaii, slow inflation over several months, perhaps accompanied by summit eruptions, tends to be followed by several days of rapid subsidence during which extrusion occurs along the rift systems that extend radially from the caldera of this shield volcano (Dvorak *et al.*, 1983). Electronic distance measurement (to 4 parts in 10^6) and levelling have been used to compute the location and volume of magma intrusions in the summit region. The tilts computed from the results on the assumption of elastic deformation were then compared with tilts measured daily and at one site by a continuously recording tiltmeter. Some of the discrepancies that emerged could be ascribed to earthquake activity, others to localized block movement superimposed on the electric displacements. In other words, they brought to light the development of structural features which were otherwise obscured by the topographic effects of eruptive sequences.

TIDE GAUGES AND SATELLITES

Terrestrial topography is generally measured by reference to sea level, and fluctuations of sea level need to be assessed lest they give misleading high or low values of vertical displacement. Again, changes in the form of the geoid, the figure of the earth represented by sea level and its imaginary construction through the continents, can give the illusion of relative movements between two points or conceal the existence of such movements.

In the Phlegraean fields, the volcanic area inland of Pozzuoli, the incidence of seismicity and slow ground motion are strongly correlated with tidal amplitude (Casertano *et al.*, 1976). All three records appear to have a period of 400-450 days, whence comes the suggestion that they are influenced by the Chandler Wobble. At any rate the coincidence is thought to indicate that sea level influences the local topography by a mechanism termed 'tidal pumping' because the volcanic deposits of the area are known to be very porous and comparable to 'a sponge dipped in water' in their rheology. Needless to say, the vertical movements could not be specified without reference to an external datum, namely the tidal gauge at Naples.

The analysis of coastal uplift and submergence has long depended on tidal data, the most familiar example being that of Fennoscandia (Lisitzin, 1974). In well managed stations the record is continuous and can therefore be purged of periodic effects (that is to say the tides themselves) to reveal whether any long-term trend is present. Three tide-gauge stations on the Atlantic coast of Canada thus show submergence at about $2 \cdot 5$-$4 \cdot 0$ mm/year between 1940 and 1970 (Grant, 1980), a rate which tallies with archaeological evidence of 2–3 mm/year. A rise of about 1 mm/year is widely accepted as the current global average, although random oscillations of ± 30 mm about this trend apparently occur randomly on a scale measured in decades and are attributable to variations in temperature and salinity. A secular trend in land height which differs from the tidal record by more than 30 mm during the period of measurement is deemed to invite the suspicion of having a tectonic explanation.

Tidal data are cheap to obtain because gauges are in existence in many parts of the world and their maintenance is an international responsibility. Nevertheless the effective period spanned is short. The Amsterdam tide gauge was set up as long ago as 1682, but as it lies in an area suspected of subsidence for tectonic reasons as well as through sediment compaction, its value for the study of crustal movements is small. The British Isles boast an excellent tide-gauge network but the length and quality of the record is inadequate for assessing the relative importance of changes in land and sea levels (Valentin, 1953; Kelsey, 1972; Rossiter, 1972; see Chapter 9). The greatest scope for tidal analyses at present is found in seas such as the Mediterranean where the record may be short but the tidal range is small and there are sheltered bays in which storm effects can be neglected. Yet even here one has to avoid sites where human activity has obscured the final pattern. At Venice (Fig. 59), for example, a rise of mean sea level of 6 mm/year in the last 20 years is due largely to subsidence provoked by the withdrawal of groundwater for the city, and both dredging and embanking have enhanced tidal amplitude (Pirazzoli, 1973).

Once adequately documented, tidal gauges can provide useful checks on elevation in areas subject to seismic disturbance. As with geodetic data their value is especially obvious in the search for premonitory movements and more generally in testing seismic models which involve dilatancy. On the west coast of North America a dozen gauges are located near major

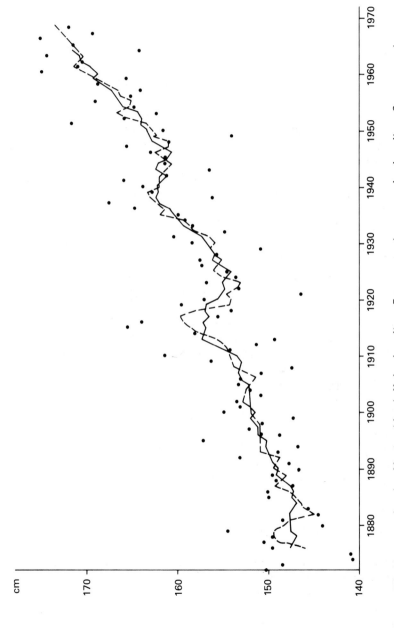

Figure 59. Mean sea level at Venice (dots). Unbroken line: 5-year running mean; broken line: 9-year running mean. After Pirazzoli (1973).

active strike-slip faults for which mean sea level has a standard deviation of about 3 cm (1·5 cm if the sea levels are checked against gauges which are not close to faults) and can thus detect movements of a few centimetres (Wyss, 1975).

Quite apart from its bearing on the tidal record, the changing form of the geoid is an important clue to the viscosity of the earth's mantle. Satellite altimetry has already proved its worth in establishing the general form of the modern geoid. Where the GEOS 1 satellite (launched in 1965) obtained a precision of ±0·6 m for a footprint 15 km across (Fig. 60), with SEASAT I (launched in 1978) the uncertainty was reduced to 0·1 m (Melchior, 1978). In a recent survey of the Philippine Sea an r.m.s. residual of 46 cm yielded a map of the geoid with a contour interval of 1 m (Horai, 1982).

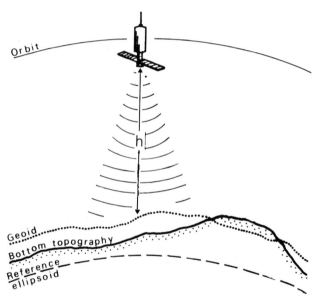

Figure 60. Satellite altimetry depends on the travel time of a radar pulse reflected from a relatively narrow footprint (shown by lines) on the sea surface.

The rapidity with which satellite altimetry can be repeated makes possible almost continuous monitoring of the geoid. LAGEOS, an aluminium satellite 60 cm in diameter with over 400 corner-cube reflectors, was launched in 1976. Tracked by laser ranging, it has already revealed changes in the earth's polar flattening. As its orbit, in common with that of other satellites, is largely governed by the earth's gravity field, any acceleration of the node of the orbit (that is, where the plane of the eclipse intersects the orbit of the satellite) indicates a decrease in the oblateness of the earth

(Kaula, 1983). The LAGEOS data, calculated on a time scale also based on Lunar Laser Ranging, give a value for $\delta\omega/\omega$ of $7\cdot1\pm0\cdot6\times10^{-11}$ years^{-1}, in excellent agreement with historical records of ancient eclipses which yield $6\cdot9\pm2\cdot6\times10^{-11}$ year^{-1}. The conversion of the satellite data to viscosity requires modelling of the proposed cause of the flattening and of the slowing down in rotation, the most probable being mass readjustment linked to crustal rebound after the late Pleistocene ice sheets had melted (Wu and Peltier, 1984). Progress in the study of recent deformation evidently needs to be complemented by improvements in Pleistocene chronology.

It must not be overlooked that the gravity map is in itself of significance in the study of recent deformation, since it permits submarine topography to be mapped and hence structural patterns to be elucidated in areas hitherto poorly mapped bathymetrically (Horai, 1982). Gravity can also be used to supplement geodetic levelling in areas suspected of active deformation, such as Iceland (Torge, 1981). Its potential as a device for continuous monitoring is currently limited by the need for elaborate environmental corrections, but one can safely look forward to increased instrumental stability, and hence to even greater advantages over ground-based topography resurvey, in that the observations can be made with a portable instrument without the need for a link to a reference level or another instrument at the time.

Most momentous of all are the measurements of interplate displacement that satellite laser ranging (SLR) is beginning to supply. Preliminary results obtained from SLR facilities in various areas (Fig. 61) are generally consistent with the findings of geophysics (Smith *et al.*, 1979). The Atlantic, for example, is widening by about $1\cdot5\pm0\cdot5$ cm/year. The agreement is gratifying; but, as often, it is in the areas of disagreement that the findings may prove most fruitful. Why, for example, is Australia apparently drifting northeastwards at 7 cm/year? And why does the rate of slip on the San Andreas Fault emerge from satellite ranging as 9 ± 3 cm/year, whereas geodetic data indicate a value of about 3 cm/year? As the two sets of measurements were not made from the same locations, one possible explanation is that the displacements picked up by the geodetic data are only those occurring near the fault, whereas the motion indicated by the satellite observations is taken up by deformation elsewhere. Another is that the short-term rates, as in the Zagros (p. 47), are higher than the long-term rates. A similar discrepancy arises in the Alpide Belt when slip rates derived from evidence from the seafloor are compared with the motion indicated by earthquakes with $M_s > 5\cdot5$ for the period 1910–70 and by that inferred from the historical record (North, 1974; Jackson and McKenzie, 1984, p. 246).

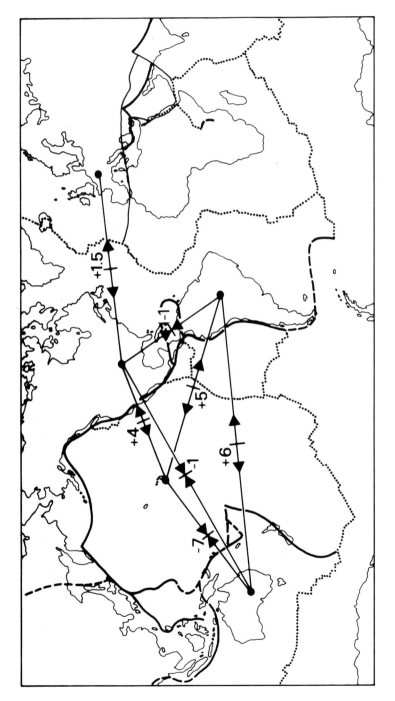

Figure 61. Intercontinental displacements determined by very long baseline interferometry (VLBI) and satellite laser ranging (SLR). Plates as in Fig. 4; displacements after Henbest (1984) by permission. © 1984 New Scientist.

Rapid replication ensures adequate statistical levels. Between March 1982 and August 1984 the SLR system operated by the Japanese Hydrographic Department at Simosato recorded 79 279 returns for 310 passes made by LAGEOS and a further 187 502 returns from two other satellites, STARLETTE and Beacon-C. Doppler satellite tracking is being used in similar fashion, for example in studies of the Earth's rotation and for navigation. By 1980 station co-ordinates could be determined by this method in the Navy Navigation Satellite System (or Transit system) to an accuracy of about 1 m. Indeed, lithospheric plate movements are seen as a possible complication once 10 cm accuracy is approached because the reference tracking stations will thereby be significantly displaced (Robbins, 1980).

Doppler determinations need to be anchored to a terrestrial reference system in order to satisfy geodetic needs. One view is that they will provide second order points within a first-order network of satellite laser points. A higher (zero) order of points would be provided by laser measurements on lunar targets. All the points would be labelled with a date and a velocity vector (Bender and Silverberg, 1975; Robbins, 1980).

The first-order system is likely to include some points fixed by very long baseline interferometry (VLBI) (Shapiro, 1983). This technique exploits the differences in the time at which a signal from a distant quasar is received at two radio telescopes, and a precision of ± 2 cm is claimed for trans-Pacific VLBI baselines of about 8000 km (K. Kasahara, personal communication). Satellites can provide radio signals which are about 10^6 times stronger than those from extragalactic radio sources and the receiving equipment is correspondingly simplified, but uncertainty over the satellite orbits still eclipses this advantage. If the last 20 years are any guide, however, the problems will soon be overcome.

STRESS MEASUREMENT

The crustal stresses responsible for deformation can sometimes be measured directly. The lapse of time enters the discussion here in two ways: first, by way of the dimensions (MLT^{-1}) implicit in any analyses of forces and, second, through the comparison of stress directions and magnitudes at successive periods.

Palaeostresses are reconstructed from geological evidence. The attempt is usually doomed to failure when, besides anisotropic and inhomogeneous rocks, one is faced with the outcome of a complex history (Ramsay, 1976). Late Quaternary and Holocene deposits are sometimes relatively homogeneous. More important, in favourable circumstances they reflect a single stress regime or at worst a sequence of regimes which have something interesting in common. For cxample, folding may have occurred at intervals in response to repeated motion on a single, major buried fault.

But, besides the distorting effects sometimes produced by pre-existing structures, partial survival or exposure of the evidence can mean that the extension carefully documented by slickenside measurements is in fact a geometrical effect of reverse faulting, or that thrusting stems from rotation of blocks on listric normal faults (Jackson *et al.*, 1982*b*). In short, however subtle the mathematical analysis of the field data, the resulting pattern of finite strain may bear little resemblance to the palaeostress field (Carey, 1979).

When it comes to measurements of the present-day stress field, the three-dimensional geometry of the measurement site is more readily assessed, but the need remains to differentiate between the current tectonic factor and pre-existing (or remanent) stresses (Ranalli, 1975). The measurements, which can be performed more than 2 km below the surface, generally vary with depth at any one location. They are influenced by topography and structure, both of which can distort the stress field, by erosion, which can give rise to large horizontal stresses, and by human activity, notably tunnelling.

The fact that the *in situ* measurements are very local in their import (Fig. 62) would not be a drawback if they were cheap and quick enough for local variations as well as averages to emerge. But the techniques on which they depend are slow and costly and, besides the many errors to which they are prone, the strain readings cannot readily be translated into

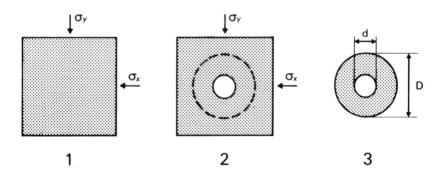

Figure 62. Principle of the in situ *stress relief method.* (1) *Original stress field;* (2) *hole bored to accommodate deformation gauge;* (3) *hole overcored to relieve stresses in the rock around the borehole and thus permit absolute stress in the rock to be calculated. Ratio* D:d *must be selected judiciously to ensure stress relief without secondary effects from heat etc.*

stresses by virtue of the complex behaviour displayed by many rock types, especially outside the laboratory (Grob *et al.*, 1975).

Two main techniques are used nowadays for the purpose. Overcoring consists of drilling a hole at a point where faults and joints are absent and installing three strain gauges perpendicular to each other at the base and on the side of the hole. An annular hole of larger radius is then drilled; this overcoring, it is assumed, relieves the stresses in the block to which the gauges have been attached, and the change in the readings gives a measure of the original state of stress. The hole can only be drilled to 1 m and for deep readings the test must therefore be carried out in mines. The hydrofracturing technique depends on pumping fluid into a part of a borehole which has been isolated from the rest and establishing from the sudden fall in pressure when a fracture has occurred. This pressure and that required to keep the fracture open are used to calculate the minimum and maximum horizontal principal stresses on the assumption that the fracture is vertical and that it fractures in pure tension (Turcotte and Schubert, 1982).

The remanent stress can be estimated by making *in situ* stress measurements on rocks of different age in the region under review. A recent study of central Europe yielded stress values for rocks ranging from PreCambrian to Tertiary which lay within a range of 20° in azimuth and 15% in amount (Greiner and Illies, 1977). The implication must be that, in comparison with the applied tectonic stress, the residual effect is not significant. An alternative method is double overcoring, where a second hole is drilled within the core to detect any additional change in stress level. In the central European study mentioned above, double overcoring showed that the measured stresses were predominantly of active tectonic origin.

In Iceland, initial stress measurements made by the overcoring method were so shallow, with the deepest at 50 m, that they were probably influenced by temperature effects and topography as much as deep-seated tectonic forces. In one instance the measurement site was only 60 m from a mountain face 100 m above the valley floor. The results indicated compression. Later studies suggested any such tendency was not in any case necessarily representative of conditions at depth. Hydrofracturing was carried out to depths of some 400 m in two boreholes on the flanks of the continuation of the Mid-Atlantic Ridge. It revealed not only that stress, and in particular the orientation of the principal stress, was dependent on depth, but also that deep-seated extension was overlain by a zone of lateral compression. Nevertheless, the principal stress direction was oriented approximately perpendicular to the axial rift zone in question (Haimson and Voight, 1977).

A recent innovation should remove some of the obstacles of cost and slowness. In the 1960s it was found at the Nevada Test Site that boreholes were elongated in a direction that was controlled by the local stress field. In 1983 it was shown that in several oil fields the elongation was, as one might have expected, in the direction of the least compressive stress, and that hydraulic fracturing data were in agreement with these results. Oriented callipers and ultrasonic acoustic devices are now being used to exploit the data available in many wells. The prospect is offered of unlimited information at a great range of depths, though without the control on the factors at issue provided by overcoring and hydraulic fracturing. Deep Sea Drilling Project (DSDP) sites are beginning to yield *in situ* stress data for the oceanic crust. At Site 504B, for example, south of the Costa Rica Rift

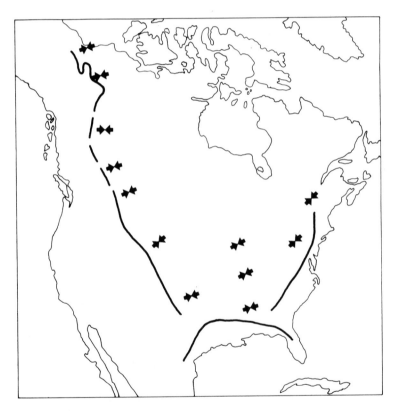

Figure 63. *Mid-continent stress province of North America (bold lines) and generalized direction of greater horizonal principal stress derived from breakouts in oil wells (arrows: each pair represents 3-30 measurements). After Gough et al. (1983) with permission. © 1983 Macmillan Journals Ltd.*

on the Nazca Plate (see Fig. 4), measurements by televiewer were made at a depth of 1000–1350 m and indicated that the average maximum horizontal compressional stress was oriented N 20 ± 16° E, consistent with focal mechanism solutions for local earthquakes (Zoback, 1983; Newmark *et al.*, 1984).

The ultimate purpose of stress measurements is usually to identify areas with relatively constant directions and magnitudes of the horizontal principal stress. Unlike methods which rely on seismicity, and are thus most profitable at plate margins, *in situ* techniques are especially valuable for analysing intraplate stress fields, which are important clues to the nature of the forces responsible for plate movement. Breakouts in oil wells show that one such province embraces the North America continent east of the Rockies with the exception of the Appalachians and the Gulf of Mexico province. The greater horizontal principal stress (S_h) is oriented NE–SW within this mid-continent stress province (Fig. 63), whereas in the adjoining Alaskan Province it is oriented NNW–SSE. It is ascribed to basal drag on the plate, although the origin of the drag is not clear (Gough *et al.*, 1983). In Australia, over 1500 *in situ* measurements which were made between 1957 and 1979 in 50 locations at depths of between 1·5 m and 1·5 km indicate a predominant E–W direction for the major principal stress (Denham *et al.*, 1980). The result conflicts with the N–S orientation suggested by crude plate-drift models and thus prompts a search for mechanisms not simply related to plate velocity, such as small-scale cross-convection.

FAULT-PLANE SOLUTIONS

In situ stress data are usually evaluated in conjunction with any pertinent evidence for recent ground deformation and the stress orientations derived from focal-mechanism solutions for local earthquakes.

Mechanism is used by some authors in the general sense of 'how the earthquake has been produced': for example, Richter (1958, p. 754) includes under it entries on compression, elastic rebound, faulting and tectonic earthquakes. What is intended here is the more restricted sense of focal mechanism, that is the kind of motion indicated by the event, such as normal, reverse, strike-slip or a combination of strike-slip with vertical displacement (Fig. 64). The answer may lead to the underlying process once it is combined with evidence from other localities and other sources of data, but it is initially limited in its import to one event and focus or, in the case of aftershock data, to one relatively short sequence of related events and one fault plane or other source of seismic energy. In short it is parochial.

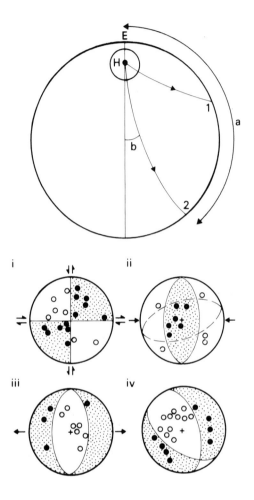

Figure 64. The focal sphere is shown surrounding the hypocentre (H). Note also the epicentre (E), the path of two earthquake waves (1 and 2), the epicentral distance (a) and the corresponding angle of emergence (b). The intersection of the seismic rays with the focal sphere permits the polarity observed at a station such as 2 in the figure to be plotted. It is usual to do so on either a stereographic or an equal-area projection of the lower hemisphere. Based on account in Kasahara (1981). The lower figures are idealized focal plane solutions. (i) Strike-slip; (ii) dip-slip reverse, (iii) dip-slip normal, (iv) normal with a small strike-slip component. Note that the data for (ii) are insufficient to constrain the solution unambiguously. The alternative (broken line) also indicates reverse faulting but on an ENE-WSW strike. At the nodal lines separating the compressive and dilatational sectors the amplitude of the initial P-pulse tends towards zero. Open circles = dilatational arrivals; solid circles and stipple = compression.

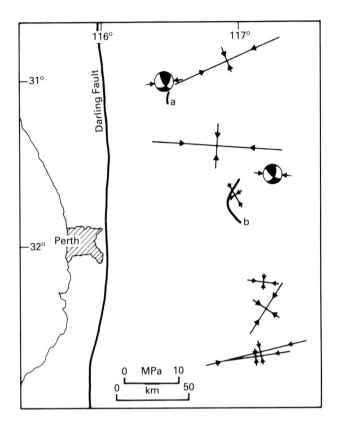

Figure 65. Stress measurements near Perth, Western Australia. (a) the fault scarp produced by the Calingiri earthquake (1970) and (b) the scarp produced by the Meckering earthquake (1968). The corresponding focal mechanisms show compressive sectors in black and the P axis by arrows. The arrowed lines represent the in situ *maximum and minimum stresses relative to the MPa scale below. After Denham* et al. *(1980).*

At Meckering, in Western Australia, stress measurement by overcoring at depths of less than 10 m has been undertaken near the epicentre of a large earthquake with the specific aim of comparing the results with those obtained from surface faulting and from analysis of the Meckering (1968) and Calingiri (1970) earthquakes (Fig. 65). The *in situ* data indicated a regional compressive stress bearing N 77° E; the earthquakes indicated pressure axes bearing N 91° E and N 102° E respectively, a level of agreement which is regarded as good. The magnitude of the observed stresses along a 200 km N-S traverse passing through Meckering was between 2 and 25 MPa; intraplate stress calculated on the assumption that

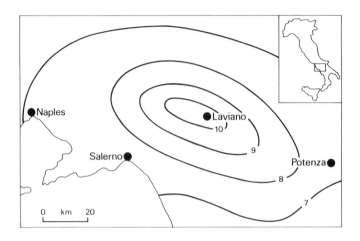

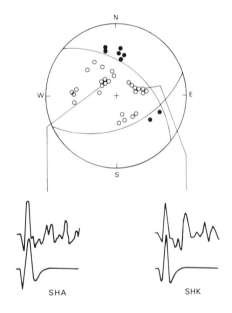

Figure 66. Location of the Italian earthquake of 23 November 1980 and intensity contours. The fault-plane solution is for the main shock (lower focal hemisphere). Dilatational arrivals shown by open circles and compressional arrivals by solid circles. Synthetic waveform modelling, shown here for only two stations, calculated for a point-source double couple. After Deschamps and King (1983); Westaway and Jackson (1984) later reported a normal fault associated with a surface break 10 km long. Note the slight first motion at station SHK, close to what would characterize a nodal arrival.

plates are driven by mid-oceanic divergence is 10–30 MPa. In other words, stress evaluation yields numerical results which can be compared with the predictions of geophysical models with regard to azimuth and force.

This is not to suggest that focal mechanisms invariable supply unambiguous directional data. When a fault-generated earthquake takes place the first P (primary) body wave to be recorded will be compressive or extensional according to the position of the instrument in relation to the seismic focus. A plot of these observations on a suitable projection will reveal a pattern of compressional and extensional zones which reflects the nature of the fault movement. The usual procedure is to plot the data on a lower hemisphere stereographic projection (as in Fig. 64). The quality of the available records and the skill of the person reading them may be inadequate for the quadrants to be readily determined. Moreover the P readings refer only to the radial displacements of the focal sphere, the imaginary surface surrounding the earthquake focus which serves as a guide to the regional displacements (Fig. 64); for tangential displacements one requires data on S (secondary) waves, and these are more difficult to read. In consequence focal mechanisms based on P waves are often ambiguous as regards which of the two nodal planes is the postulated fault plane and which the auxiliary plane orthogonal to it (Scheidegger, 1963). The problem is discussed in Chapter 9.

The solution bears on conditions at the depth of the hypocentre, a fact often obscured by maps which moor the stereographic projections to points on the ground surface without any indication of the relevant depth, and sometimes overlooked by those comparing seismic data from depths of ~ 10 km with stress measurements made at ~ 10 m.

Major earthquakes are usually located teleseismically, that is to say by reference to stations in the global network, and their position can be in error by 10 or even 30 km mainly because of uncertainty over arrival times and the velocities at which the waves are transmitted to the various stations. As many earthquakes do not break the surface, deformation is only occasionally a useful guide to the location of the epicentre. Intensity derived from reports of damage and ground motion experienced by people rather than instruments (Table 3) are not a dependable substitute. Initial accounts of the damage produced by the Italian earthquake of 23 November 1980 indicated an epicentre near the coast between Naples and Salerno, whereas it lay inland beneath the village of Laviano (Fig. 66) (Deschamps and King, 1983). The lack of clear evidence for surface faulting, compounded by poor weather and a snow cover in the higher parts of the affected area, meant that the correct location was much delayed.

One may be able to relocate the main shock by monitoring aftershocks with the help of portable seismographs (Figs 67 and 68), and to eliminate

Table 3

a) Modified Mercalli Intensity Scale of 1931 (1956 version)

I. Not felt. Marginal and long-period effects of large earthquakes.

II. Felt by persons at rest, on upper floors or favourably placed.

III. Felt indoors. Hanging objects swing. Vibration like passing of light trucks. Duration estimated. Many not be recognized as an earthquake.

IV. Hanging objects swing. Vibration like passing of heavy trucks; or sensation of a jolt like a heavy ball striking the walls. Standing motor cars rock. Windows, dishes, doors rattle. Glasses clink. Crockery clashes. In the upper range of IV wooden walls and frame creak.

V. Felt outdoors; direction estimated. Sleepers wakened, Liquids disturbed, some spilled. Small unstable objects displaced or upset. Doors swing, close, open. Shutters, pictures move. Pendulum clocks stop, start, change rate.

VI. Felt by all. Many frightened and run outdoors. Persons walk unsteadily. Windows, dishes, glassware broken. Knick-knacks, books etc. off shelves. Pictures off walls. Furniture moved or overturned. Weak plaster and masonry D cracked. Small bells ring (church, school). Trees, bushes shaken.

VII. Difficult to stand. Noticed by drivers of motor cars. Hanging objects quiver. Furniture broken. Damage to masonry D, including cracks. Weak chimneys broken at roof line. Fall of plaster, loose bricks, stones, tiles, cornices (also unbraced parapets and architectural ornaments). Some cracks in masonry C. Waves on ponds; water turbid with mud. Small slides and caving in along sand or gravel banks. Large bells ring. Concrete irrigation ditches damaged.

VIII. Steering of motor cars affected. Damage to masonry C; partial collapse. Some damage to masonry B; none to masonry A. Fall of stucco and some masonry walls. Twisting, fall of chimneys, factory stacks, monuments, towers, elevated tanks. Frame houses moved on foundations if not bolted down; loose panel walls thrown out. Decayed piling broken off. Branches broken from trees. Changes in flow or temperature of springs and wells. Cracks in wet ground and on steep slopes.

IX. General panic. Masonry D destroyed; masonry C heavily damaged, sometimes with complete collapse; masonry B seriously damaged. (General damage to foundations.) Frame structures, if not bolted, shifted off foundations. Frames racked. Serious damage to reservoirs. Underground pipes broken. Conspicuous cracks in ground. In alluviated areas sand and mud ejected, earthquake fountains, sand craters.

X. Most masonry and frame structures destroyed with their foundations. Some well built wooden structures and bridges destroyed. Serious damage to dams, dikes, embankments. Large landslides. Water thrown on banks of canals, rivers, lakes etc. Sand mud shifted horizontally on beaches and flat land. Rails bent slightly.

XI. Rails bent greatly. Underground pipelines completely out of service.

XII. Damage nearly total. Large rock masses displaced. Lines of sight and level distorted. Objects thrown into the air.

Masonry A = good or especially reinforced; B = good; C = ordinary; D = weak.

b) *The Rossi-Forel Scale*

I. *Microseismic shock.* Recorded by a single seismograph or by seismographs of the same model, but not by several seismographs of different kinds; the shock felt by an experienced observer.

II. *Extremely feeble shock.* Recorded by several seismographs of different kinds; felt by a small number of persons at rest.

III. *Very feeble shock.* Felt by several persons at rest; strong enough for the direction or duration to be appreciable.

IV. *Feeble shock.* Felt by persons in motion; disturbance of movable objects, doors, windows, cracking of ceilings.

V. *Shock of moderate intensity.* Felt generally by everyone; disturbance of furniture, beds etc., ringing of some bells.

VI. *Fairly strong shock.* General awakening of those asleep; general ringing of bells; oscillation of chandeliers; stopping of clocks; visible agitation of trees and shrubs; some startled persons leaving their dwellings.

VII. *Strong shock.* Overthrow of movable objects; fall of plaster; ringing of church bells; general panic; without damage to buildings.

VIII. *Very strong shock.* Fall of chimneys; cracks in the walls of buildings.

IX. *Extremely strong shock.* Partial or total destruction of some buildings.

X. *Shock of extreme intensity.* Great disaster; ruins; disturbance of the strata, fissures in the ground, rock falls from mountains.

The two scales correspond roughly as follows:

MM	I	II	III	IV	V	VI	VII	VIII	IX	X-XII
RF	I	I-II	III	IV-V	V-VI	VI-VII	VIII-	VIII+ to IX-	IX+	X

modified after Richter (1958)

Geomorphological phenomena are first noticed at M.M. VII and are significant only at M.M. IX (R.F. X).

some of the least reliable stations by comparing observed and calculated arrival times. Depths are more elusive especially where shallow earthquakes are concerned. A nuclear explosion in the Aleutians on 29 October 1965 was located teleseismically 26 km NNW of the shot point and at a depth of 76 km instead of 660 m (Auden, 1970). The main reason is that stations are rarely close enough to the epicentre. The problem is greatly reduced if one can rely on local records obtained at distances from the epicentre which are less than the focal depth (Jackson *et al.*, 1982a; Molnar and Chen, 1982).

The orientation of the nodal planes is likely to have an uncertainty of ±10°. In addition, the proposed fault plane need not be simply related to the regional stress field. Many earthquakes, especially if they are shallow, occur on existing faults (McKenzie, 1969). Evidently an existing fault plane

Figure 67. Portable seismograph being set up to record aftershocks.

Figure 68. Aftershock record. Each tick on the lines parallel to the long side represents 1 s. The record is on a sheet of paper originally attached to a rotating drum, so that each line, read from left to right, is continued on the left by the next line below it. The black diagonal lines merely link up the spaces left between successive minutes of record on each line. Note the large number of aftershocks and the variations in their form (and hence their origin). The record is for Verchia, southern Italy, 8 December 1980. Courtesy of Dr G. C. P. King.

at a large angle to the maximum principal stress will tend not to be exploited, but the uncertainty remains considerable. Aftershock data may vary in this respect, and also as regards mechanism, to the point of confusion, which is hardly a surprise in a zone which has been broken up by successive shocks. The influence of pre-existing structures on seismic activity shows how easily the distinction between current and 'geological' movements can break down. The fact that several indicators—say fault plane solutions and fold patterns—give a similar orientation does not necessarily reinforce the case, for both could be the slaves of an earlier, influential trend. Only by identifying and keeping separate the successive phases of motion can the historical effect be measured.

Chapter Six

GEOLOGICAL
IMPLICATIONS

'I don't ask for much, but what I get should be of very good quality'.
Prayer in New Yorker

Our growing understanding of the pattern and tempo of recent crustal deformation already colours the way we view geological history as a whole. It is also having some effect on the way geology is investigated, although more as a matter of emphasis than anything far-reaching or unsettling.

Take that old catch phrase 'The present is the key to the past'. Although it retains some didactic value, the wisdom it enshrines grows daily more difficult to spell out. An obvious complication is that organic evolution brings with it irreversible changes and hence irreconcilable differences between ancient and modern geological agencies. For example, the emergence of plant life meant among other things that the landforms of temperate areas would develop quite differently from those in arid lands and, by the same token, that some ancient 'desert' landforms could have formed in areas that lacked a vegetal cover but were not climatically dry (Brooks, 1926, pp. 199, 248). But that is no longer all. Support is growing rapidly for the view that uniformitarianism, the doctrine that came to replace catastrophic interpretations of Earth history by the middle of the nineteenth century, need not be taken to mean anything more than that natural laws have always prevailed (Gould, 1965). It is naive—so the argument runs—to equate uniformitarianism with uniformity of rate. Periodic catastrophic events may in fact have more effect than long periods of gradual change (Ager, 1973).

This dispute is not a fruitful one. Lyell, the standard-bearer of uniformitarianism, was careful to confine his argument to 'the permanency of the laws of nature' (Lyell, 1837) and to exclude only imaginary mechanisms from the acceptable repertoire of geological agencies. Those

121

who eagerly raise the catastrophist standard today by invoking asteroid impact to explain the extinction of the dinosaurs cannot escape the Lyellian embrace, as the Earth boasts several meteorite craters whose nature is accepted by the most conventional of earth scientists. Indeed there is already a school which speaks of actualistic catastrophism (Hsü, 1983).

The real weakness of uniformitarianism lies in its appeal to modern processes whose nature and effectiveness we do not fully understand. In short, the key is often a blank. An important if partial exception is the work of rivers. Even by Lyell's day a large body of data had been collected largely by civil engineers concerned with the upkeep of canals and the control of rivers which gave a reasonable picture of the rate and rhythm of sediment transport by several major watercourses. The annual sediment load of the Ganges provided Lyell with one of his most vivid illustrations of river action, just as his own observations on Vesuvius and in Georgia served to demonstrate the rapidity with which gullies could be gouged out (Lyell, cited by Chorley et al., 1964, pp. 162-4, 189). Uniformitarian arguments made it possible to calculate the time taken to construct the delta of the Rhine in Lake Constance and thence the duration of postglacial times (Charlesworth, 1957, p. 1522). They also underlay estimates of the age of the Earth obtained by comparing the dissolved load of rivers with the ocean's salinity.

The fact that Joly thereby obtained an age for the Earth of 81 million years when later calculations gave a value of over 1 500 million years is merely a warning against the use of inadequate data. To extrapolate a trend from a period of record spanning a few decades into the whole of the Earth's history is to perform the sort of calculation that allowed Mark Twain in *Life on the Mississippi* to conclude that the Mississippi, which was allegedly shortening its course by cutting across the necks of some of its meanders to the tune of several metres a year, presumably once stuck out into the Gulf of Mexico like a fishing rod.

There was little temptation to do likewise with crustal movements because no comparable set of data was to be had. To be sure, the topographic effects produced by the Indian earthquake of 1897 soon demonstrated the importance of seismic deformation, but in the absence of information on the frequency of such earthquakes the cumulative effect could hardly be assessed. Scandinavia and Pozzuoli* were viewed as puzzles of local

---

*At a meeting devoted to non-elastic processes in the mantle, Sir Harold Jeffreys suggested that every department of geophysics should display a photograph of the 'temple' of Serapis 'as a reminder that the phenomena of change of level are not so simple as most discussions assume' (Crittenden, 1967*b*, p. 279). His primary concern was to remind the audience that subsidence and uplift there had not been in response to changes in surface load.

significance; the desire to cast out catastrophist, doctrine, coupled with the search for a story swamped in a rain of sediment gentle as talcum powder, did the rest. The lack of progress in documenting earth movements may be seen in the authoritative claim made in 1975 that too little was known about rates of deformation for the corresponding forces to be estimated (Price, 1975). It is no wonder that textbooks of geodynamics published in recent years still confine their discussion of vertical crustal movements to Scandinavia, and that in one explicitly devoted to the interrelationship between tectonics and landforms only eight of its 310 pages deal with rates of crustal deformation (Ollier, 1981). Structural geology and related aspects of historical geology and geomorphology have grown accustomed to an almost static setting. The vast aeons of time that radiometric dating revealed had as their corollary a panorama changing too slowly for the changes to be perceptible. Evidence of catastrophes helped to sustain the illusion: the catastrophes provide the wherewithal to explain cumulative change even when gradual modification has apparently left no trace in the record.

DEFORMATION AND EROSION

The relative speed with which features are produced by crustal deformation as opposed to erosion or deposition smacks of mediaeval disputation about needles and angels. But the fact remains that the relative importance of tectonics and of external agencies in the development of topography is central to theoretical geomorphology as well as to such practical matters as the use of morphological evidence in structural mapping on land, beneath the sea and on other planets.

The topic is commonly presented as a dispute between two giants of geomorphology, William Morris Davis and Walther Penck (Fig. 69). Davis — so the simplified story runs — invoked uplift as a device for initiating the cycle of erosion, and in his idealized narrative kept any further tectonic activity at bay until the cycle had run its course, whereas Penck postulated a scheme in which crustal movements and erosional processes operated simultaneously. A glance at their writings (Davis, 1909; Penck, 1953) will show that this is a caricature of complex views rich in qualification. Yet it is how the scheme is encountered by those students who encounter it at all: all too often geology texts still give pride of place to the Davisian story or at least his terminology (for example Holmes, 1965, pp. 471-475).

In consequence, field geologists will have preconceived notions about the significance of features to which tectonics and erosion are both likely

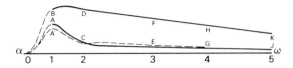

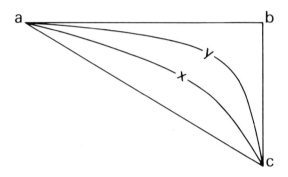

Figure 69. Tectonics and landform development. Upper: The idealized erosion or 'geographical' cycle according to Davis (1899). Relatively rapid uplift at B is followed by erosion, with relief shown by AB, CD, EF, GH and JK. Note that some incision during uplift is conceded by the broken line OA, which subsequently represents aggradation. Lower: The model of W. Penck postulates a continual interplay between erosion and tectonic change. ab indicates the effects of crustal movement and bc the effect of denudation. Their combined product is symbolized by the curves all starting at a and ending at c. After Penck (1953) with permission.

to have contributed, such as fault scarps. Alternatively they may hesitate to jump to conclusions about such evidence. How can they know whether a scarp is a true fault scarp, rather than a fault-line scarp picked out by erosion along the contact between two contrasting bodies of rock juxtaposed by faulting, especially when doing so from satellite imagery?

The evidence cited in earlier chapters shows that the Penckian view, if not his model, often comes closer to reality. Tectonic activity is not a matter of rapid regional uplift followed by stability for the million or so years apparently required for the postulated erosional cycle to run its course. But the Davisian may counter this assertion with two arguments. Features corresponding to the pattern predicted by the idealized cycle of erosion are widespread. Anyone who has read about young valleys or old ones will recognize examples without hesitation. And in any case the nub of the question is not whether uplift was effectively instantaneous but whether tectonic processes can outpace erosion long enough and widely enough for an old cycle to be wiped out and a new one initiated.

An attempt to tackle the problem was made by Schumm (1963). His approach was to collect maximum rates of uplift from various parts of the world and to compare them with maximum rates of erosion to see which set gave the higher values. The answer seemed to be unambiguous: uplift, with averages of up to 7 m/1000 years, could easily outpace erosion, with maxima of about 1 m/1000 years. Schumm concluded that time-independent slopes, as postulated by Penck on the premise that erosion and uplift could attain some kind of balance, were unlikely. Conversely, rapid uplift as proposed by Davis for starting off his cycle (and his critics, it must be repeated, do not give him adequate credit for the cautious way in which he made the proposal) appears reasonable.

The data used in the study were not truly compatible. The erosional figures represented the volume of sediment supplied by drainage basins and averaged out per unit area. Uplift rates commonly refer to uplands or terraces selected for the very fact that they are suspected of anomalously rapid uplift, and hence are unlikely to be regionally representative.

In any case the area encompassed by the Davisian unit, say a mature valley or a peneplain, no longer justifies treatment as a single genetic feature, and even if the Earth's crust were to remain stationary while an erosion cycle worked itself out, the climate is unlikely to prove equally obliging. The 'normal' landscape which long served as a standard of reference turns out to be the product of several contrasting climates during the Pleistocene and owes little to the agencies currently at work. Changes of sea level have added to the confusion by rejuvenating coastal rivers and ensuring that 'youthful' rivers should coexist with 'old' topographies.

In short, rather than vindicate Davis's model, advances in the study of recent deformation are helping to dig its grave. (It would be premature to speak of burial. Like uniformitarianism, the geographical cycle is persuasive and difficult to replace in the classroom.) The terminology of youth, maturity and old age will doubtless persist, because it is a convenient form of shorthand for describing a complex set of features. But there are abundant graphic and numerical alternatives (Scheidegger, 1970) to such phrases as 'steep slopes' or 'low relief'.

The advantage of these alternatives, charmless though they may appear, is that they avoid many of the assumptions inherent in the Davisian model, or, for that matter, in Penck's or in any other view of landscape viewed as an entity. The decision whether or not to derive tectonic data from the landscape is deferred. If other sources are defective, or for some other reason

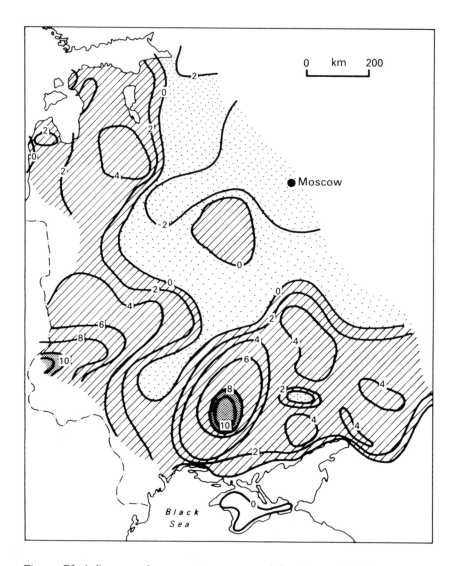

Figure 70. Influence of recent movements of the Russian Platform on river action. Lines are contours of recent movement in mm/year; the shaded area shows where the length of non-meandering rivers exceeds 50% of the total length of the rivers, supposedly because erosion is encouraged by uplift. After Mescheryakov (1967). The use of different contour intervals explains most of the differences between this map and Fig. 55. Reproduced with permission from Keter Publishing House, Jerusalem Ltd.

to be insulated from the discussion, the slopes, channels and all the other components can be inspected for evidence of deformation (Fig. 70). River terraces can thus be used in the search for warping, their sole contribution being the property of sloping down valley. Their identification as river terraces arises from a host of incidental facts, not least the observer's experience, but their origin—climatic, tectonic, man-made etc.—is omitted as far as possible from the preliminary stages.

Combining 'pure' morphological data with independent accounts of recent deformation carries the benefits one step further. Implicit in much palaeogeographic work is the assumption that massive clastic deposits imply active tectonic uplift which in turn provides a source for weathered debris. For example, the Bakhtiyari Conglomerates of Iran have long been considered the product of mountain building in the Zagros Belt. The assumption is difficult to test for reasons that include the influence of human activity on erosion rates, but there is already enough information to show that—as one would have guessed—rapid uplift cannot guarantee aggradation and aggradation does not necessarily imply tectonic activity.

Massive clastic deposits mantle the limestone uplands of Epirus (Fig. 71) and interfinger with the beds laid down in the basin of Lake

Figure 71. Late Quaternary alluvial and colluvial deposits (left foreground) at Kokkinopilos, Greece. The Louros valley is to the right (east).

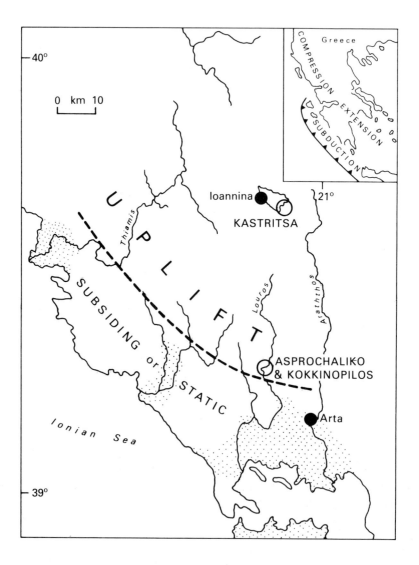

Figure 72. Effect of fault movement on river behaviour in Epirus, northwestern Greece. The broken line marks the switch from coastal subsidence to uplift inland. After Bailey and King (1985).

Ioannina about 20 000 years ago. It has recently been suggested (Bailey and King, 1985; cf. Vita-Finzi, 1978) that localized tectonic warping promoted aggradation by creating dam-like obstructions in the valleys (Fig. 72). Yet it is difficult to isolate the effect: deposition occurred in all the valleys of Epirus principally in response to increased weathering by frost and accelerated slope erosion. The process began long after deformation was already in operation and has yielded to erosion despite continuing tectonic instability. In other words, deformation locally accentuated an existing trend.

The connection between tectonics and erosion or deposition may be easier to demonstrate in reverse. In his discussion of the Davisian model, Schumm (1963) observed that the original estimates for the time it would take to wear the USA down to a peneplain (the end-product of a Davisian cycle) overlooked the isostatic uplift produced by erosion, and he substituted an estimated 33 million years for the original 10 million. The issue is most memorably illustrated (Jeffreys, 1970, p. 406) by the Arun, a river which rises in the Tibetan Plateau and crosses the Himalayas between Everest and Kanchenjunga. Granted that it is an antecedent stream, that is to say one that maintained its course as the mountain range it crosses was being elevated, and on the assumption that the crust here has a density of $2 \cdot 7$ whereas the mantle has a density of $3 \cdot 3$, erosion of a kilometre of rock from the top of the range implies the addition of $0 \cdot 7$ km below so that the net lowering is $0 \cdot 3$ km. But as erosion is largely confined to the slopes and valleys, the summits will be uplifted isostatically in response to this erosion quite apart from any relative height above the valleys they acquire in the process. There is, presumably, a comparable effect at work in areas recently subjected to severe glacial erosion.

UNCONFORMITIES AND CORRELATION

Another potential casualty of progress in the study of deformation is uncritical correlation based on eustasy: as with the Davisian cycle, 'the very geological advances which first appeared to strengthen the eustatic theory later helped to discredit it' (Chorley, 1963, p. 964). Eustasy, as promoted by Edward Suess at the turn of the century, implied continental stability, and by extension encouraged the search for erosion surfaces of continental origin which had originally been graded to specified positions of sea level. Attempts were thus made to match such surfaces in the eastern USA and the British Isles, for example, and countless studies of coastal terrace sequences relied for dating and correlation on their height even when the area of study was known to be prone to earthquakes and contained young mountain belts in the western Mediterranean. A major exponent of the tradition, F. E. Zeuner, expressed his confident opinion that the raised beaches would eventually provide a link in dating Pleistocene and Palaeolithic remains in coastal regions all over the world (Zeuner, 1958, p. 127).

Although some Quaternary geologists are still pursuing the correlation of fossil beaches on the basis of height, the majority have become chary of placing undue reliance on an item which could prove treacherous, and are turning increasingly to independent dating when time-correlation is a central concern of their researches. But the eustatic torch has not been extinguished. Stratigraphers who work in oil exploration, as we saw earlier, have maintained and perhaps even revived interest in global sea level because they believe that it is a valuable device for correlating and dating rock sequences, especially in areas that are poor in exposures and borehole records.

The key concept, as in all eustasy, is that there is a global pattern that can be identified from local sequences once anomalies and distortions have been filtered out. Or, to put it another way, that tectonics and sedimentation effects will be eclipsed by the depositional products of sea-level fluctuations. Where the age of the strata in question is based on detailed fossil analysis and the coastal advances and retreats are read from the sediments, one can accept that broadly synchronous shifts in widely separated areas are at least as likely to be produced by sea level as by simultaneous movements of the land. Indeed, the weight of opinion—doubtless still impressed by the ease with which glacial advances and retreats can explain falls and rises of the sea, and unimpressed by all attempts to explain why earth movements in different parts of the globe should be synchronised—is inclined to view the bath as stable and the bathwater as fickle.

The revived interest in eustatic correlation has had its major impact in sequences which have not been sampled directly and hence whose age is inferred on the basis of unconformities and their supposed correspondence with sea-level episodes. However objectively this work is done, there remains the risk that the number of events will be adjusted to expectations and that breaks open to a variety of interpretations will be ascribed axiomatically to sea-level changes. In some recent studies the unconformities are traced offshore to depths so great that, unless massive crustal movements are invoked, they must have formed below sea level (Vail and Hardenbol, 1979).

But before becoming too enmeshed in the issue of correlation, we should consider the idea of base level control in its own right. The concept is ingrained in physical geology. First formally stated by J. W. Powell (1875), it holds that erosion cannot operate below the local or regional limit imposed by trunk streams, hard rock outcrops and ultimately the sea (see also Greenwood, 1857; Chorley *et al.*, 1964, pp. 382ff.). If we picture the area immediately adjacent to the coast there can be no argument: rivers that flow into the sea cannot cut down appreciably below low tide, and if there has recently been local submergence the streams will doubtless be engaged in silting up their lower reaches. But once we move inland the connection between local and general base level becomes extremely tenuous and consequently elusive when all we have to go by are a few boreholes and quarry faces. To take an extreme case, would it be obvious from a borehole at the longitude of Jericho that the Jordan was not graded to the contemporaneous sea but to the Dead Sea 390 m lower?

A related point is how far sedimentary history reflects changes (as distinct from individual positions) of sea level. The base-level thesis has as one of its corollaries the onset of sedimentation if base level rises and of erosion if it falls, and we have seen how reasonable these proposals are close to the shore. Once we move inland, however, sea-level changes could be a matter of indifference to the rivers, lakes, glaciers and lakes responsible for deposition, let alone those parts of the sea floor still submerged. The case is especially clear in the case of a fall in sea level, where one expects downcutting to work its way inland forthwith. The slowness with which this process operates in anything other than soft alluvium is illustrated by the Niagara Falls, which are thought to have taken about 12 000 years to progress some 10 km, and this with little effect on the land outside the gorge. Elsewhere, hard rock outcrops have greatly slowed down the headward retreat of waterfalls and escarpments and hence confine regrading of the channel to a narrow coastal belt.

There is little to suggest an orderly progression from erosion to aggradation, with the sea acting as regulator and unconformities as surfaces of uniform age bounding bodies of sediment, and even less when we look at the mounting evidence for crustal mobility. If the lands or tracts of sea floor are on the move, any attempt to distinguish between eustasy and tectonics will be doomed, perhaps not to failure but certainly to uncertainty.

The verdict does not amount to a retreat from global generalisations. It simply rejects the assumption that contemporaneity is implied by the presence of unconformities either at the same elevation or in similar stratigraphic positions. A good example of the need for scepticism is the widely-quoted evidence in the Persian Gulf for rises of sea level to 100 m or more above the present (Kassler, 1973). On the Arabian coast the evidence includes sabkhas now inland which are thought to represent relic arms of the sea, and a plain 80–120 m high which is considered to be of marine origin. The high stands appear consistent with the eustatic record of the Lower Pleistocene and comparisons have been drawn with the classic terrace sequence of the Mediterranean, although the possibility of some tectonic distortion is generally recognized.

An attempt to check the proposed chronology on the Saudi Arabian coast led to the conclusion that no unambiguous Pleistocene marine terraces were to be found more than 4 m above present high water. Fossiliferous deposits of Miocene age are among the features that perhaps explain earlier views: they give rise to terraces and their fossils are not surprisingly beyond the reach of the radiocarbon method (McClure and Vita-Finzi, 1982). Whatever the validity of the high sea-level stands elsewhere, they are

Figure 73. Landsat view of the Strait of Hormuz. The Musandam Peninsula of Oman (left) has undergone submergence during the Holocene through the combined effects of the postglacial marine transgression and subsidence of the land towards the northeast at a maximum rate of 8·5 mm/year. The Zagros shore of Iran (right) has experienced uplift by up to 7·0 mm/year largely through the growth of anticlines parallel to the coast.

apparently absent here and thus do not date—let alone explain—successive erosion surfaces inland. In compensation, however, the terraces that are present, being invariably lower than their synchronous counterparts on the Iranian shore, provide a valuable guide to differential tectonic movements in the Gulf. The Arabian coast emerges as far more stable than the Iranian, except on the Musandam Peninsula, where Holocene subsidence is at a rate very similar to that of the uplift of the opposing Zagros coast (Fig. 73).

One might expect shores separated by spreading axes to display analogous tectonic histories, especially in the early stages of rifting, but even there the resemblance will be confined to a narrow coastal belt. Offshore the eustatic factor, which at times is modulated by activity at mid-oceanic ridges and by other tectonic controls of oceanic capacity, presumably confers a greater uniformity on the record, but its influence will be restricted laterally to an even greater extent. There remains one further important source of tectonic change which could conceivably justify stratigraphic correlation: mountain building.

The search for orogenic phases of global significance has long been pursued, but before the days of radiometric dating the synchrony of the deformation in widely separated areas owed much to preconceptions. For instance, theories that invoked global contraction required a much higher level of global synchrony than those in which the filling and eventual compression of geosynclines played a major part. A recent version of the hypothesis which invokes intermittent contraction at the surface postulates at least 20 eras of mountain-building at intervals of about 100 million years (Lyttleton, 1982). Although the weaker zones will be the first to yield, the process is a worldwide one. Compare this picture with accounts in which the European geosynclines, destined to become Hercynian orogenic belts, were already filling as the Caledonian movements reached their climax; or the wave-like migration of orogenic phases across the Sunda area from mid-Carboniferous times through the late Tertiary (Holmes, 1965). Even now terms such as Alpine and Hercynian are sometimes used of mountain ranges or their remnants more on the basis of their trend and structural style than from evidence that they formed in response to a common driving force.

Side by side with the identification of orogenic stages or episodes, there has been a sustained effort to trace the development of individual mountain chains or complexes. Here too the narrative, being based on incomplete sections and a fragmentary record, may sometimes appear more chronologically assured than is strictly justified.

The advent of plate tectonics has not eliminated the problem: although mobility is general and circumstances conducive to mountain building

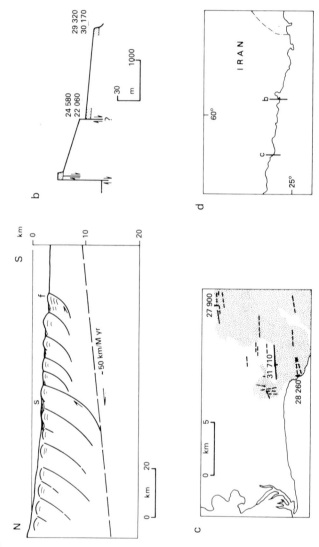

Figure 74. Southern Iran. (a) Schematic N-S section across the Makran accretionary prism showing imbricate thrust faults above subducting oceanic basement. Note frontal fold (f) and sediments trapped by ridges and then tilted (s). After White and Louden (1982), who postulate uplift on the coast (left) of about 1·5 km/million years (= an average of 1·5 mm/years). (b) Konarak peninsula, showing ^{14}C ages for the front and rear of the main marine terrace, a possible fault bounding the terrace and faults at the rear of peninsula. Uplift at 2 mm/year would explain the discrepancy between the ages for the front and rear of the terrace. (c) Faulted marine terrace (stipple) at Gocsar, showing ^{14}C ages for three locations. Heights range from 2·5 m near the coast to c. 20 m inland and demonstrate how a single surface can be disrupted by E–W faults.

After Vita-Finzi (1981). (d) Position of sections (b) and (c).

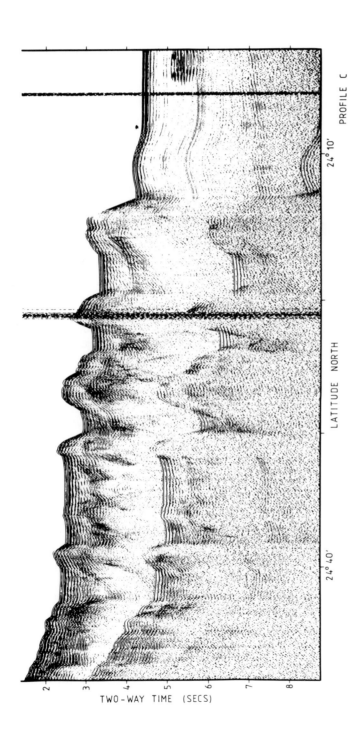

Figure 75. Continuous seismic reflection profile across the Makran continental margin off the Pakistan coast (near the bottom right-hand corner of Fig. 40(d). Vertical exaggeration at the sea floor is 7:1. Note frontal fold on the right, and interfold basins showing increase in the dip of the sediments with depth. From White and Louden (1982), courtesy of Dr R. S. White.

are widespread, major episodes of orogeny have been localised in time and space. There are mountain belts of which only the roots survive and others in the earliest stages of development. As well as being patchy the evidence is complicated even if certain major categories of mountain range, such as island arcs and cordilleran-type belts, have come to be accepted.

Information on recent movements can make a modest but still valuable contribution to the subject, notably by specifying the different modes and rates of movement during a narrowly specified period at different points along a youthful mountain belt. At a more advanced stage in the enquiry it may of course be possible to link these variations to the corresponding seismic and geophysical picture.

The coast of southern Iran includes parts of two major structural provinces, the Zagros and the Makran, separated by a fault zone which runs roughly N-S. This zone, often referred to as the Oman Line and in its southern part as the Zendan Fault, has for many years been viewed as something of a puzzle by structural geologists. First, there was the abruptness with which it separates the simple folds of the Zagros, composed of sediments ranging in age from Palaeozoic to Tertiary, from the largely Cainozoic ophiolithic melanges and turbidites of the Inner Makran. Second, the transcurrent motion on a N-S alignment that could most readily explain the break appears to have given way in the Miocene and Pliocene to E–W shortening (Shearman, 1976). The current view is that the Zagros belt represents collision between two continental areas, whereas in the Makran (Fig. 74) oceanic crust is being subducted beneath the Iranian plate (White and Klitgord, 1976; Farhoudi and Karig, 1977). Geophysical data suggest that one consequence is the development of folded sediment ridges offshore and parallel to the Makran coast which are in due course uplifted along thrust faults (Fig. 75). The rate of subduction in the Makran is put at about 50 mm/year and the angle at which it occurs is thought to be about 1° from the horizontal. There is no general agreement about the rate of convergence between eastern Arabia and Iran at the Gulf, but palaeomagnetic data from the Red Sea suggest that during the last 5 million years the Arabian peninsula has been separating from Africa at about 20 mm/year from an axis striking N 110° E, that is on an azimuth of about N 20° E (Girdler and Styles, 1978).

This general pattern of deformation accords with the recent geological record on the Zagros coast, where active folding has led to progressive tilting of strata and to uplift of beaches cut into anticlines still in process of development. At Khomsokh, west of Bandar Abbas, the modern wave-cut platform has a seaward dip of 2° or less. Late Pleistocene marine sediments now 20 m above sea level rest on an ancient platform which dips about 10°, and the underlying Tertiary beds, truncated by the platform, have

Figure 76. Quaternary marine deposits at Khomsokh, Iran, resting unconformably on Tertiary beds. Note that the older beds dip more steeply seawards (towards the left front) than the younger ones.

Figure 77. Marine terraces cut into an anticline in the coastal Zagros at Tujak, Iran. The figures are on a beach 11·8 m high and dated by ^{14}C to c. 3000 years. A terrace 28·6 m high and dated to c. 6500 years can be seen on the skyline to the right.

Figure 78. Anticline at Tujak at the rear of the marine terraces shown in Fig. 77. Its continued growth is responsible for uplift of the terraces.

dips of about 20° (Fig. 76). The position of this sequence on the limb of an anticline parallel to the coast suggests that tilting is a product of progressive anticlinal growth. Further south, at Tujak, uplift has been sufficiently rapid for Holocene terraces to be raised by as much as 28·6 m above present-day high water (Figs 77 and 78). There is widespread agreement that sea level during the last 7000 years did not depart by more than 10 m from its present position, but its precise position at particular times is uncertain. The uplift graphs in Fig. 79 have been adjusted on the assumption that there has been a gradual if decelerating rise of sea level during the Holocene. Some believe there was a transgression above modern datum followed by a slight fall. The first scheme is derived from the average values obtained from different parts of the globe; the second is based on glacioeustatic considerations and depends on position relative to the Pleistocene ice caps (Vita-Finzi, 1979*b*, 1982). The resulting averages differ little because the eustatic factor here is small compared to the tectonic.

The calculation was performed for three anticlines on the Zagros coast and their uplift rates were converted to rates of horizontal shortening by using the approximation

$$\Delta y = \Delta X \times 2\pi/\omega \times (s/L)^2$$

where ΔX is the horizontal displacement, Δy the vertical displacement of a point P on a fold at a distance s along the fold limb from one of the troughs, ω is the maximum dip of the fold limbs in radians, and L is the length of the fold surface between two troughs with s/L small (Fig. 80). Holocene shortening with an average resultant of about 26 mm/year on a

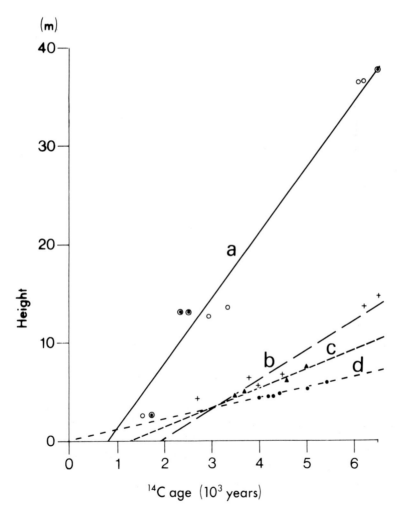

Figure 79. Uplift rates for fossil beaches at Tujak (a), in Bahrein (based on data from Doornkamp et al., 1980) (b), at two sites near Bandar Abbas (c) and on Qatar (d). Heights adjusted for eustatic effects.

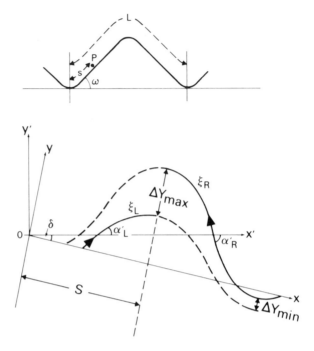

Figure 80. Fold analysis. Upper: Symmetrical curves of intrinsic sine form used for converting uplift at P to horizontal shortening. On the assumption that folding takes place over a slide surface (décollement) and that s is short in comparison with L, one can use the approximation $\Delta y = \Delta X \times 2\pi/\omega \times (s/L)^2$. Lower: Symmetrical, sinusoidal fold tilted through an angle δ which can be fitted to partially exposed or eroded fold structures using measured dips and the pairing of some of the strata. After Mann and Vita-Finzi (1982).

bearing of N 36° E was indicated, in good agreement with the results obtained in the Red Sea and consistent with the proposed mechanism. It shows how the rigidity of lithospheric plates can be tested if one has independent evidence of rates both at the trailing and at the leading edges; and it suggests that most of the folding in the Zagros belt is confined to its southwest margin, as any additional deformation inland would make the 'effect' unreasonably larger than the 'cause'. As it happens, there is already wide acceptance of the idea that folding of the Zagros is serial in character, with deformation largely confined to the frontal fold on the Gulf coast or slightly offshore.

A reverse fault on the Zendan lineament indicates 3·5 m of movement 7000–1250 years ago on a plane dipping 32° SW (Fig. 14). The graph of uplift against age at Tujak, one of the coastal Zagros sections, again

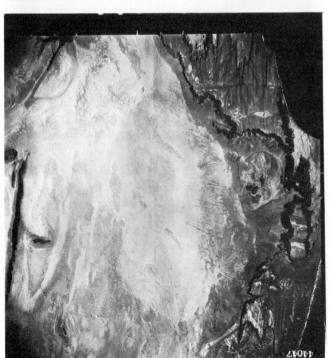

Figure 81. Normal faults in Pleistocene marine terraces near Gavatre on the Makran coast of Iran (stereopair). The lower edge of each photograph represents about 12 km.

suggests there has been no movement for the last thousand years or so (Fig. 79). Although variations in the rate of deformation are to be expected and indeed assumed in any discussion of the topic, the Holocene evidence from the Zagros and its eastern margin is a further reminder that averages get more representative the longer the period they encompass but thereby obscure extreme values which could be no less informative. And the apparent hiatus of the last millennium shows how misleading a picture of recent deformation in the Zagros one would have gained from evidence confined to that quiet period.

In the Iranian Makran the coastal belt includes many striking instances of late Quaternary faulting (Fig. 81). Average Holocene rates of uplift increase towards the Pakistan border in the east to a maximum of about 2 mm/year. Radiocarbon dating shows that marine platforms which formed during the last 35 000 years have been disrupted by predominantly normal faulting. The presence of successive terraces at individual sections supports the view that uplift was seismic (Page *et al.*, 1979) and it is interesting to note that subduction at a horizontal rate of 50 mm/year with uplifts of 2 mm/year implies that any subduction was at an angle of about 2°.

The dip of the sediment trapped between the folds off the Makran coast indicates progressive tilting towards the coast (Fig. 75). The pattern of movement accords with the view that folding characterizes the seaward part of a wedge of sediment which is being scraped off the oceanic Arabian plate as it drives under Eurasia (White and Louden, 1982). The process is still going on. A recent age for the tilted beds makes sense in an area such as this where occasional flash floods sweep large volumes of sediment from the marl and clay coastal hills out to sea.

As in the Zagros, the observations merely show that the postulated mechanism gives reasonable answers: one would expect only part of the convergence to be taken up by coastal deformation and the rest to result in thrusting (Figs 13 and 14) and other processes inland. But fresh issues are raised by the exercise, notably the eastward increase in rate, which could reflect variations in the rate of convergence, and the appearance of 'extensional' structures in an area subject to prolonged compressive forces.

In due course such findings will make it possible to calculate strain rates, a vital step towards an explanatory mechanism. As things stand, the fact that fault displacement is easier to trace than tilting or folding is reflected in the relative prominence given to the time dimension in discussions of fault rocks. The Alpine Fault of New Zealand is an interesting example. It is a zone about 1 km wide and includes various rocks whose character owes much to fault movement — hence the term 'fault rocks'. Movement amounts to 480 km of dextral displacement since the Jurassic, although in the late Miocene the transcurrent trend gave way to movement with a component of reverse slip. One interpretation of the various rock types sees each of three textural groups as the product of successive phases of movement in the late Jurassic–early Cretaceous, the late Tertiary and the Quaternary (Adams, 1981; Sibson *et al.*, 1981; see also Jeffreys, 1942). Another view is that most of the crucial fabrics formed in the late Miocene, the different types reflecting depth in the fault zone rather than time of origin. Potassium/argon (K/Ar) dating indicates vertical fault movement of about 2·5–5 mm/year for roughly the period of 2–5 million years ago, and 7–4 mm/year for the last 700 000 years. Shear heating was probably at its most intense after the changeover from transcurrent to oblique movement as higher stresses would have been needed to maintain slip. With a shear resistance of about 1 kbar, a horizontal component of 12–24 mm/year for 5 million years would produce an adequate thermal effect. In fact movement may well have exceeded 50 mm/year.

Information on the progress of folding in due course will make calculations of this kind possible for anticlines and synclines. Although the mathematical analysis of folds has made great advances, field data still tend to be drawn up using either the Busk technique, which relies on arcs of circles, or freehand interpolation. In consequence, the results of experimental and mathematical analysis of alternative mechanisms are not readily tested in the field.

The shortening required by the Zagros coastal folds discussed earlier was calculated with an approximation which embodied the assumption that the folds were of intrinsic sine form. A more complex method which proposes a sinusoidal form for the folds gives computed curves which are in good agreement with the observed dips of synclines and anticlines at Tujak (Fig. 80) (Mann and Vita-Finzi, 1982). Elsewhere it may be sufficient to consider rates of tilting and uplift before taking the crucial step from geological description to geodynamic explanation (Lewis, 1971; Wellman, 1971).

Chapter Seven

GEOPHYSICAL APPLICATIONS

Some facts can be explained in several ways even when there is actually only one correct explanation.

Lucretius*

Advances in our understanding of recent deformation are of potential benefit to many topics in geophysics besides the mechanics of folding and faulting. Alfred Wegener (1929) opened his review of the evidence for continental drift with geodetic arguments, for these represented a direct test of the theory. In a later chapter he referred to the recent uplift of Scandinavia as a product of isostasy and its corollary that the 'crustal underlayer' displays some degree of fluidity. Isostatic effects remain important clues to the viscosity of the mantle, whose nature is evidently of critical importance to any discussion of global convection models especially in the context of drift. Structural geologists continue to derive valuable insights into the nature of the Earth's crust from the pattern of recent folds and faults. And seismologists—as the next chapter shows—are increasingly hungry for information on the history and former distribution of earthquakes as well as on the ground deformation accompanying current seismic activity.

The connection between recent movements and geophysical inference is not always simple. For instance uplift rates of 5–10 mm/year for the Nanga Parbat region are required in order to explain unexpectedly young fission-track ages because the uplift would lead to active erosion and cooling: fission-track dates record the time when the rock passed through its closure temperature, which is controlled by cooling rate. The centre of maximum

---

*Sunt aliquot quoque res quarum unam dicere causam/non satis est, verum pluris, unde una tamen sit (Lucretius, *de rerum natura* VI, 703-704).

uplift is found to lie at the point of convergence between the Hindu Kush, the Karakoram and the Himalayan ranges for reasons which await elucidation (Zeitler *et al.*, 1982). But the literature is still dominated by the issue of mantle viscosity and postglacial isostatic readjustment, and the application of deformation rates to other problems remains in its infancy.

ISOSTASY AND MANTLE VISCOSITY

If Wegener viewed the relative fluidity of the subcrust as an argument for horizontal as well as vertical movement of continental blocks, Daly (1934) saw that, by matching the history of unloading with the observed pattern of uplift, one would be able to obtain numerical estimates of the 'fluidity'. For the analysis to succeed, one must evidently be sure that the uplift is indeed the product of unloading. The case of Lake Bonneville is instructive.

In 1890 G. K. Gilbert surveyed the remains of Lake Bonneville, a lake which had occupied an area of about 50 000 km^2 in the Great Basin of the USA during the Pleistocene (Figs 82 and 83). According to him the lake had undergone two cycles of filling and desiccation, and the resulting shorelines were warped upwards in such a way that the maximum deflection from the horizontal was where the water had attained its maximum depth of 335 m and exerted the greatest load.

A few authors have proposed alternative explanations calling for deformation which, though ultimately responsible for the lake (and doubtless influenced by its presence), was not initiated by filling and voiding of the basin. But the consensus is that any such effect has been swamped by the isostatic phenomenon (Gilbert, 1890; Crittenden, 1967a). An objection often made to secular upwarping as the source of deformation is that it would have had to stop while the shorelines developed and then gone on to affect an area which precisely matched that occupied by the lake. But the correspondence is hardly unexpected if the basin was in fact the product of secular—if spasmodic—structural forces. At all events faulting is not at present considered significant. A series of fault scarps has come to light in the southern part of the basin near Lund, but Gilbert had ignored this area because he found the field evidence unclear. Recent movement on the Wasatch Fault which bounds the basin on the east (Fig. 84) if anything has countered the observed uplift as the downthrow is to the west (Crittenden, 1963).

Besides confirming and in some respects refining Gilbert's field observations later workers have derived values for the viscosity of the mantle from the history of uplift. The first step is to allow for the elastic

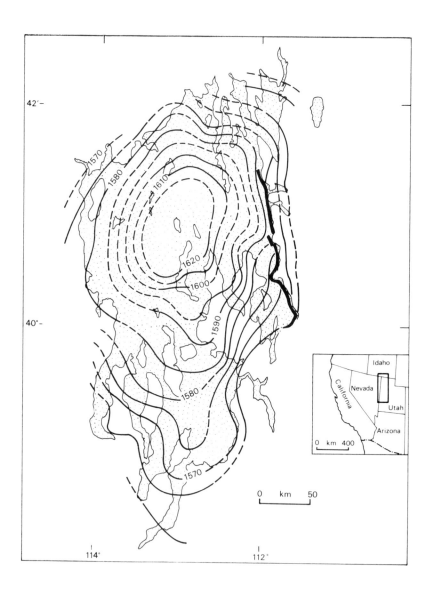

Figure 82. Deformation of the Bonneville shoreline. Isolines converted from the original values in feet to rounded metric equivalents. Area of lake at Bonneville shoreline stippled. After Crittenden (1967a). The heavy line denotes the Wasatch Fault.

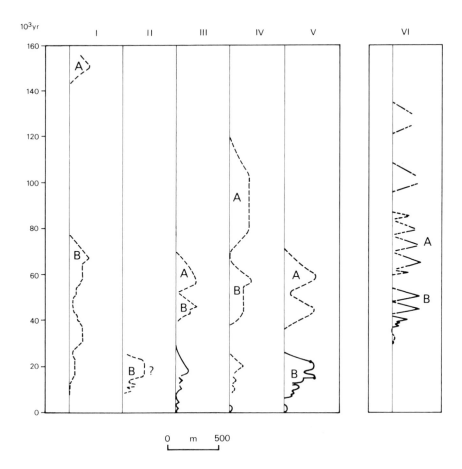

Figure 83. Different views of late Quaternary chronology of Lake Bonneville. Age is indicated by the vertical axis except for VI which is a purely relative chronology. V shows variation of load with time. Water depth is shown by the curves with a common scale at the base of the diagram. Broken lines indicate uncertainty. A: Alpine stage; B: Bonneville shoreline. The dots on curve V are ^{14}C-dated shorelines. I-IV after Crittenden (1967b), V after Crittenden (1963), VI after Morrison (1965a), with permission. © The American Geophysical Union.

compression produced by a water load of about $10 \cdot 16 \times 10^{18}$ g at the maximum extent of the lake. On the assumption that the load is laterally infinite, and that the crust is here about 50 km thick and has a Young's modulus of about 7×10^{11} dyn/cm^2 ($= 70$ GPa) and a Poisson's ratio of $0 \cdot 25$, the total vertical shortening is 178 cm. An alternative calculation for a load of the diameter of Lake Bonneville, namely 200 km, gives 152 cm

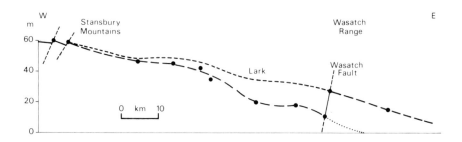

Figure 84. Displacement at the eastern margin of Lake Bonneville produced by the Wasatch fault. After Crittenden (1967a). Large dashes: observed shorelines. The upper dashed line on the left shows the original position of the shoreline if the absolute motion on the Wasatch fault is all down on the west; the dotted line at lower right shows the original position if absolute motion on the fault is all up on the east.

at the centre of the load (Crittenden, 1967b). The effect tends to be ignored because its size is close to the limits of error of the estimated shoreline elevations.

In calculating the viscosity of the mantle, the assumption has usually been that it flows as a Newtonian fluid and that its viscosity is uniform. The time taken for the deviation from isostatic equilibrium to fall to $1/e$ of its initial value—the relaxation time or T_r—is taken to be proportional to viscosity and inversely proportional to the linear dimension of the load that was removed in order to facilitate analysis in a basin with irregular shores and depths. Crittenden evaluated the average depth of water within a series of circles drawn over the basin for computing the load. A mantle density (ϱ) of $3 \cdot 25$, in agreement with seismic data for the area, makes it possible to calculate the progress of isostatic compensation. For example, as the amount of depression produced by a depth of water h is h/ϱ, 280 m of water would give a deflection of 86 m, and subsequent uplift by 64 m would represent 74% adjustment. The quality of the computation is of course improved by substituting water volume for depth and allowing for the shear strength of the crust. The figure most generally accepted is about 75% of compensation.

The deflection produced by a given load will amount to $1 - 1/e$ (or 63%) in the period defined above as T_r. An early attempt to calculate T_r suggested that it lay somewhere between 4000 and 10 000 years. The adoption of a revised chronology for Lake Bonneville in 1967 showed that T_r was probably 4000 years or less. According to the equation used by Vening Meinesz to calculate mantle viscosity in Fennoscandia, the effective viscosity η (poise)$= \varrho g t_r/2f$, where T_r is in seconds and f is a factor which incorporates the horizontal dimensions of the load.* The result, expressed cautiously to indicate that it is very much an approximation, is $\sim \times 10^{21}$ poises.

The significance of the relaxation time in the calculation is obvious. Yet the chronology of the lake remains uncertain. The model currently favoured was formulated in 1962, but the radiometric dates cited in its support (Thurber, 1972)[†] refer only to the latest part of the record (Fig. 83, Curve V) and stratigraphic work published in 1965 (Morrison, 1965a) shows a more complicated set of major lake oscillations than that to be found in the geophysical literature. In addition, the contribution made by folding in response to subsurface faulting awaits investigation. Comparison with the results obtained in the glaciated shield areas is thus premature. One authority has gone so far as to state that the Basin and Range region in which Lake Bonneville lay is highly anomalous and the structure of the Earth there is probably not comparable with that of the Scandinavian or Canadian shields (Walcott, 1973). The possibility of variations in viscosity with depth and even of dislocation motion (Smith, 1974) would of course invalidate the simple relationship between viscosity and the dimensions of the lake on which the linear factor depends. But what is known of the geometry and chronology of recent deformation already permits alternative models to be tested with some degree of conviction.

Until a few years ago the Fennoscandian postglacial record was the main source of information on mantle viscosity, as its glacial and coastal history had received prolonged attention and yielded a persuasively simple picture. Vigorous field research and the enthusiastic application of ^{14}C dating have now rendered the American evidence accessible to the geophysicist; but Fennoscandia retains the advantage of having boasted an ice sheet more regular in shape than the American (Lliboutry, 1971) and hence more readily analysed in dynamic terms. Indeed, ^{14}C and amino-acid dating of shells contradicts views held since 1943 about the location of major ice divides in the Hudson's Bay area, and only now is there talk of bringing into discussion of the Laurentide Ice Sheet a sensitive clue to postglacial rebound hitherto neglected, namely the tilt displayed by glacial lake shorelines (Peltier and Andrews, 1983). Nevertheless, although Fennoscandia was the subject of pioneering studies in which the counting of laminated lake deposits or varves permitted some coastal deposits to

---

*$f = \pi/lm/(l^2 m)^{1/2}$ where l and m are the horizontal dimensions of the load in cm. For the northern part of Lake Bonneville $f = \pi/1 \cdot 4 \times 10^7$ cm. Cathles (1975, appendix VI) uses an alternative factor which yields a slightly higher viscosity than that obtained by Crittenden, namely $1 \cdot 7 \times 10^{21}$ poise.

[†]Kaufman and Broecker (1965) suggest that ^{230}Th dating may in due course extend the record back to 200 000 years ago.

be dated very precisely, many shorelines conventionally ascribed to isostatic uplift have been identified solely on the basis of their form or at best by reference to associated archaeological remains. The history of the Baltic Ice Lake during the latter part of the last glacial (Weichselian) glaciation thus remains 'indeterminate at many points', and the ^{14}C dates obtained for the Ancylus transgression in southern Finland have been 'very variable' (Eronen, 1983, p. 204).

But on the whole these are questions of detail and the viscosities derived from the early work have been little affected by advances in beach chronology. Where opinion is beginning to experience a major conceptual shift is over the relative importance of faulting. There is growing field evidence in Fennoscandia for faulting in the course of Holocene uplift, and the suggestion has been made that the glacio-isostatic factor gave way to tectonic control 2000–3000 years ago (Mörner, 1980). Uplift in response

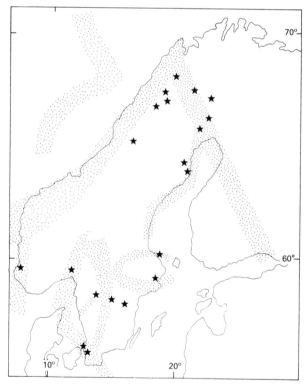

Figure 85. Recent tectonic activity in Fennoscandia. Stars indicate fault movement dating from the last 13 000 years; stipple represents seismotectonic zones derived from instrumental data and tectonic structures. Adapted from Mörner (1980) and Stephansson and Carlsson (1980).

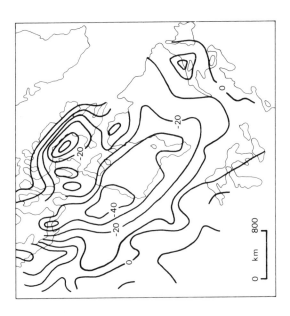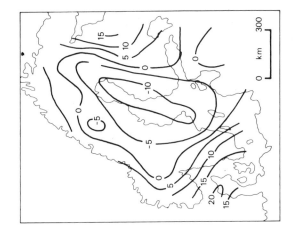

Figure 86. Free air gravity anomalies in mGal over Hudson Bay (left) and Gulf of Bothnia (right). The data are in agreement with the viscosity of the lower mantle derived from analysis of orbital data for the laser geodynamics satellite LAGEOS. After Peltier (1983) with permission. © 1983 Macmillan Journals Ltd.

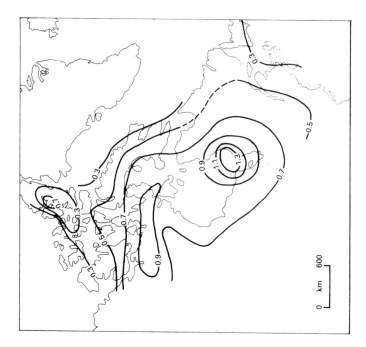

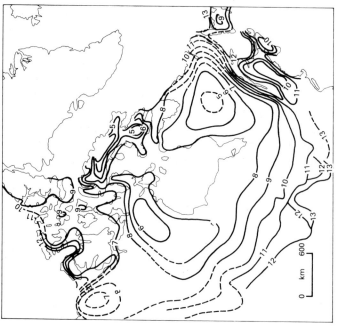

Figure 87. Left: *generalized pattern of deglaciation in northern and eastern America.* Right: *rate of present uplift in metres per 100 years. After Andrews (1970).*

to fracture would of course explain immediately why there are discrete beaches at different heights as it is likely to cause intermittent uplift (Jeffreys, 1975). On the other hand, for it to explain the oldest beaches the faults would need to have dominated uplift for the last 11 000 years. Unfortunately glacial scouring has removed the sediments that might have permitted late Pleistocene and Holocene faults to be measured and dated, and one is left with the knowledge that postglacial fault displacements of up to 100 m have been recorded in southern Sweden and that the seismic map for the last five centuries suggests that certain ancient structural lineaments have remained active (Fig. 85).

In addition, a simple glacio-eustatic interpretation of the Fennoscandian record faces growing evidence for uplift throughout the Mesozoic and Cainozoic, albeit with occasional bouts of subsidence (Lyustikh, 1960). Though not inconsistent with Pleistocene rebound, any such long-term tendency would evidently complicate numerical analysis of the recent uplift data. It would also help to account for the third problem, namely that there is little correlation between the uplift pattern and the gravity anomalies (Fig. 86) that have been used for nearly half a century to establish the extent of isostatic disequilibrium and thence the amount of uplift still to be expected (Jeffreys, 1970, p. 429; Walcott, 1973, p. 20). The 'fault' objection in fact gains support from the apparent match between gravity anomalies and rock structure. Making allowances for topography improves the correspondence between present-day uplift and gravity (Wu and Peltier, 1983), but the structural and long-term evidence remains tantalisingly vague.

The North American record can be expected to go on improving in areal detail as well as in chronological precision. The emphasis on ^{14}C dates is matched by appropriate adjustments to convert the ages to calendar years (Cathles, 1975, p. 202) so that the history of ice removal and of uplift (Fig. 87) can be compared dependably with uplift and sea-level sequences from other areas. The work is being conducted side by side with modelling of the uplift to be expected with ice sheets of different configurations and with a mantle whose viscosity either is uniform or changes with depth. Moreover, the discussion embraces deformation of the sea floor beyond the ice margin in response to loading and unloading both by the ice and by the oceans.

In other words, the real potential of glacio-isostatic research will come from its combination with studies of relative sea level suitably co-ordinated by means of radiometric ages. Preliminary work using a Laurentide deglaciation model derived in part from ^{14}C dating of terminal moraines (ICE-2) suggests that the deformation history and the map of gravity anomalies are reasonably consistent with a uniform mantle viscosity of about

10^{22} poise (10^{21} Pa s). The absence of an increase in viscosity below the discontinuity at a depth of 670 km indicated by seismic data removes one argument for confining convection to the upper mantle (Peltier, 1981; Peltier and Andrews, 1983).

Recent deformation has fuelled attempts to calculate mantle viscosity in areas far removed from glaciation. On the Huon Peninsula of Papua New Guinea there is a spectacular staircase of coral reefs which is thought to represent the combined effects of eustatic sea levels and (at least so far as the last 120 000 years are concerned) uplift at about 2 mm/year. Stillstands, transgressions and regressions identified in the coral reefs have been used

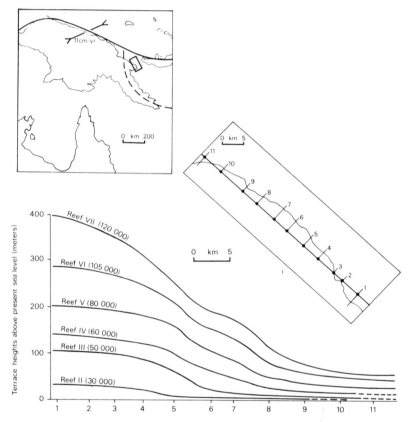

Figure 88. Profiles of elevated marine terraces on the Huon peninsula of Papua New Guinea based on the 11 sections shown in the inset, corrected for sea-level fluctuations, and dated by U-series and [14]*C. The inset map shows location of study area in possible subsidiary of the Australian plate which is converging with the West Pacific Plate as shown by the arrow. After Chappell (1974), with permission. © 1974 The American Geophysical Union.*

to construct a sea-level curve (Chappell, 1974). These is evidence of intermittent uplift, including closely spaced wave-cut notches, and the area's high seismicity would seem to point to discontinuous deformation. Even so, the assumption that the average rate of uplift is reasonably uniform gives results that agree with findings obtained elsewhere. The terraces are flexed along the coast, with the greatest uplift in the southeast (Fig. 88). Uplift between 120 000 and 80 000 years ago was apparently more rapid than it has been during the last 80 000 years but the pattern of deformation changed little.

Instead of the simple cycle of loading followed by unloading, that is usually postulated for areas undergoing glacio-isostatic uplift, the peninsula has experienced pushing and bending by virtue of its position near the junction of two (or possibly three) converging lithospheric plates. To judge from Bouguer gravity anomalies as low as -160 mGal it is also subject to isostatic uplift. Nevertheless, as at Bonneville, the procedure was to compare the age and height of successive terraces with the values predicted by different models of the crust and mantle. Given an elastic lithosphere the main point at issue is how mantle viscosity varies with depth. As in some previous studies, the calculation was facilitated by viewing the mantle as a series of parallel layers of different viscosity, although the limited length of the terraces meant that the results could bear only on the upper 400 km. The analysis pointed to an elastic crust ($\mu = 10^{11}$ dyn/cm^2) overlying a zone of low kinematic viscosity ($\sim 2 \times 10^{21}$ stokes) between 80 and 220 km below which viscosity appeared to rise to 10^{22}. Beneath Fennoscandia the low viscosity zone extends down to 400 km, a difference which could stem from the thermal effects of subduction under the Huon area.

Other sources of estimates include the figure of the Earth, temperature profiles, seismic data and experiments on plausible rock constituents (Wyllie, 1971). Of these only the first concerns us by virtue of the changes discussed in Chapter 5. Two published values for the average viscosity of the mantle derived from the figure of the Earth are 10^{26} and 5×10^{22} poises. The evidence provided by postglacial unloading emerges as superior for two reasons: it is endowed (at any rate potentially) with a record spanning several thousand years, and it bears on different parts of the globe. But the two sources are seen to be interdependent once we consider the non-tidal component of the acceleration of the Earth's rotation, as the mass loading produced by the ice sheets has to be allowed for when analysing the progress of this acceleration (Wu and Peltier, 1983).

Detailed analysis of closely dated fossil shorelines promises to reveal the form of the geoid at successive time intervals and thus to extend the brief period spanned by satellite data into the Holocene and possibly beyond. The enterprise is of significance in two ways: a good grasp of the form of the sea-level surface at successive periods will make it possible to correct the height of former shorelines and thus to measure local deformation, and the changing form of the geoid will serve to test various models of crustal distortion in response to glacial history. The fact that much of the data on geoidal history comes from the very shorelines it is designed to correct or analyse does not invalidate the exercise: the risk of circular argument is side-stepped by restricting height comparison to features of the same ^{14}C age.

One such study (Faure *et al.*, 1980) reconstructed waterplanes for 6500, 5500 and 1800 years ago perpendicular to the coast of Senegal by obtaining samples both on the coast and up the estuary of the Senegal River, formerly an arm of the sea: Holocene marine shells are found over 120 km inland. Evidently care had to be taken to ensure that the dates were intimately related to a specific position of the sea. Shallow water and intertidal molluscs or peat were accordingly selected for dating. The results showed little change in the form of the geoid since 6500 BP and any seaward tilting measured less than 1 m where some models had predicted depression of

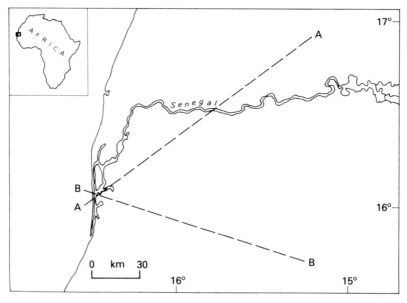

Figure 89. Survey lines (AA-BB) used by Faure et al. (1980) to reconstruct the form of the geoid inland from the Senegal estuary 6500 ± 200, 5500 ± 200 and 1800 ± 200 years BP using ^{14}C-dated sea-level data. The results show that any tilt over a distance of 120 km did not exceed 1 m, whereas some mathematical models (e.g. Clark, 1980) predict differential emergence in the region. Reproduced with permission. ©1980 The American Association for the Advancement of Science.

the ocean floor under the load of meltwater and uplift of the continental mass (Fig. 89).

Attempts are now being made to extend this approach to the Earth as a whole. Preliminary results show that even in the course of the last 6000 years the geoid has been continually deforming (Newman *et al.*, 1980). The news may be bad for those who have been seeking a single, global, eustatic sea-level curve, but it is excellent for those who hope to derive information on the viscosity of the mantle—and indeed the elasticity of the crust—in areas other than those dominated by glaciers and glacial lakes. The resulting pattern could well shed light on any changes in mass direction associated with convection, a useful complement to the contemporary map yielded by gravity surveys.

TECTONIC HISTORY

The nature and effects of the interaction between crustal plates is another topic that will increasingly rely on recent deformational history for raw data and novel applications. The classic studies that led to general acceptance of plate tectonics combined two distinct sets of data: the 'instantaneous' picture supplied by first-motion analysis of earthquakes endowed with instrumental records, and the long-term view supplied by palaeomagnetic data, palaeontological correlation and submarine geometry. The emphasis on convergence or divergence rates expressed in centimetres per year tended to obscure the lack of information on the recent past. A glance at any of the key areas, such as the North Atlantic, will show that the familiar spreading value of 3 cm/year for the Juan de Fuca Ridge, for example, depends on a palaeomagnetic time scale of which the latest subdivision spans about 700 000 years. More usually, reliance on several anomalies away from the axes of spreading gives rates averaged over 3 million years or more (Minster and Jordan, 1978).

Attempts to subdivide the palaeomagnetic scale and to identify the corresponding units on the sea floor or on land hitherto have failed to produce consistent results. Variations in spreading rate, which would evidently be of value to analysis of the driving mechanism, can only be inferred over long periods. In the Gulf of Aden, for example, two phases of sea-floor spreading have been identified of which the first lasted from about 30 to 15 million years ago and the second includes the last 5 million years (Girdler and Styles, 1978). These units cannot yet be subdivided, to the detriment of their geodynamic interpretation.

Direct chronicling of plate interaction will supply some of the missing narrative. The major fault zone which separates the Arabian and African plates along the Dead Sea Rift, and the San Andreas Fault system, which lies between the Pacific and North American plates, are instances of relatively simple boundaries on land which furnish a detailed record of relative crustal displacements. Where the boundary is convergent, as in Andean or island-arc settings, or where it is beneath the sea, the record of recent deformation will be less readily translated into plate geometry, although, as we have seen with regard to the Zagros, it is possible to derive slip values from folds provided that they are well exposed and their uplift history is datable.

The North Anatolian Fault zone (Fig. 12) extends for a distance of about 1200 km and is usually viewed as the boundary between the Black Sea and Anatolian plates. Displaced river channels, offset outcrops and linear scarps point to lateral strike–slip motion which appears to have been the rule throughout the Neogene and Quaternary (Allen, 1975; Jackson and McKenzie, 1984). Historical movement associated with earthquakes is reported along different stretches of the zone and the fault-plane solutions are in good agreement with the inferred sense of motion. There is also some evidence of creep.

Many of the recent displacements follow Holocene and earlier lineaments. The potential ambiguity of the field evidence is illustrated by reports of left-lateral motion in the Pontus Formation, which is of Upper Miocene–Lower Pleistocene age (Hancock and Barka, 1980). The evidence consists of joints and faults. In the few cases where the two indicators intersected, those indicative of left-lateral shear were cut by the right-lateral group. The investigators were careful to observe that left-lateral shear need not indicate reversal of motion as it could arise from local mechanical effects where faults overlap. In view of the dominance of joints over faults in the field data (85:15) and the small size of the faults (generally less than 20 m² in area and with displacements smaller than 2 m), the latter interpretation is more probable, especially as no comparable observations have been made outside the Neogene basins in question.

More problematic are the reports from the central and northern Aegean that the extensional regime dominant in the area since the Neogene was briefly interrupted by compressional episodes, one in the Upper Miocene–Lower Pliocene and the other in the Lower Quaternary. The duration of the episodes in itself poses problems, as mantle convection is too ponderous to accommodate such brief reversals. A possible solution was to be found in changes in the motion at the boundaries of the region, such as the North Anatolian Fault.

We have seen that there is some evidence for such an event on the fault in the Upper Miocene–Lower Pleistocene, and also that it is open to an alternative interpretation. The Aegean data appear more unambiguous.

Figure 90. Normal fault in lake beds at Çendik, near Burdur (Turkey). Note that the downthrown block (left) consists of colluvium.

The proposed sequence is based on detailed fieldwork in Locris-Euboea, the Gulf of Corinth and the islands of Kos, Samos, Chios and Rhodes (Angelier, 1979; Mercier *et al.*, 1979). But a determined attempt to find comparable evidence in western Turkey, close to Kos, Samos and Chios, failed despite the presence of excellent exposures in the Neogene. Whereas the Greek islands display steep folds and reverse faults, western Turkey provided abundant evidence of extensional tectonics (Figs 91 and 92), but only small amplitude, long-wavelength folding and tilting of a kind commonly associated with normal faulting (Jackson *et al.*, 1982*b*). The sole exception

Figure 91. Misleadingly fresh-looking fault plane exposed by erosion of overlying sediments near Manisa, Turkey.

was a structure near Burdur which may be a thrust but has remained enigmatic since its first description in 1956.

The simplest explanation for the conflicting evidence is that the structures ascribed to shortening were produced by rotation of older fault planes and strata in the course of listric faulting. For instance, steep antithetic normal faults that have been rotated by movement on neighbouring major faults may resemble reverse faults. Moreover, small faults in superficial sediments need have little regional significance. Just as reverse faulting at depth can produce extension at the surface accompanied by normal faulting, so can normal faulting produce minor reverse faulting at the surface. And many instances of puzzling uplift can be explained by elevation of the footwall of normal faults.

In any case, the compressional features postulated for the Aegean may well have appeared regionally synchronous merely because the age of each local item of evidence could not be expressed more closely than 'Lower Pleistocene' or 'Mio-Lower Pliocene'. If we accept a round figure of

Figure 92. Slickensides showing normal movement on fault surface in massive limestone near Burdur, Turkey. All that can be said about the corresponding phase of movement is that it predates the undisturbed Quaternary beds which covered the fault surface.

1-9 million years for the duration of the Quaternary in the Aegean, its lower third amounts to about 600 000 years; the even looser confines of the Mio–Lower Pliocene span about 28 million years. Moreover, the attribution of deformed beds to these time-periods is uncertain. The marine faunas used for dating fossil beaches are often composed of species still living in the area and reflect local conditions rather than age. Inland, terraces and soils are dated by matching phases of development with episodes in the glacial record of the Alps without any evidence to support the correlation and in the absence of a reliable chronology for Alpine glacial history.

The fact of extension says little about the underlying mechanism, which need not be the same in the Aegean as in other areas, such as the Basin and Range province of the USA, which are being stretched (Jackson and

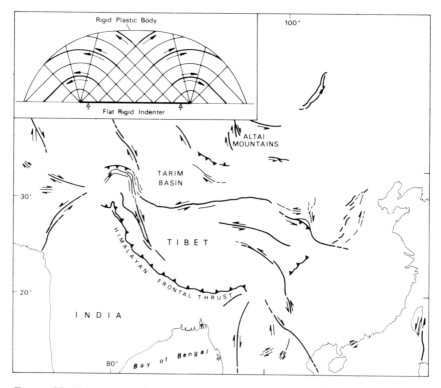

Figure 93. Major recent faults in Asia whose recency and sense of motion has been inferred from surface faulting during earthquakes, analysis of air photographs or fault-plane solutions. Inset: pattern of slip lines produced at the yield point when a plastic medium is indented by a flat rigid die. Based on Molnar and Tapponnier (1975) with permission. ©1975 The American Association for the Advancement of Science.

McKenzie, 1983). But by removing a spurious puzzle it demonstrates that the normal faulting indicated by fault-plane solutions has predominated in western Turkey throughout the Pliocene and the Quaternary, and thus provides a time scale against which plate kinematics can be unfolded.

Convergence on a grand scale is epitomized by the Tibetan Plateau (Fig. 93). Most authors now take it for granted that the plateau was raised to its present level some 5 km above the sea in response to the collision of India with Asia, but there is much disagreement over the precise mechanism. For instance, there are those who believe that underthrusting by India was responsible, whereas others ascribe uplift largely to shortening and thickening of the Tibetan crust (Tapponnier et al., 1981).

During the last few years, field studies combined with analysis of satellite imagery have shown that during the upper Quaternary and perhaps longer the dominant trend in southern Tibet has been E–W extension. The N–S normal faults that provide its clearest indication are perpendicular to the folds that betray N–S shortening during the Mesozoic and Tertiary. The E–W folds appear to be relatively ancient because they have been eroded into 'mature' landforms, are locally covered by recent volcanoes and have been truncated by faults which strike at high angles to the fold trends. Some of the faults also cut features thought to date from the last glacial period in the area and are taken to be younger than late Pleistocene in age; others are provisionally ascribed to known historical earthquakes. Strike-slip faults serve to link the normal faults, and some strike-slip movement has occurred on normal fault systems arranged en échelon. There is no evidence of reverse faulting or folding in the Quaternary and all the field evidence points to extension. But radiometric dating should in due course reveal to us whether there has been any shift in the main locus of deformation, how the normal faults are related in time to structural trends outside Tibet and how fast strike-slip movement has progressed during the Holocene.

For, according to the model (Molnar and Tapponnier, 1978), convergence between India and Eurasia, rather than taking the form of crustal shortening in Tibet, causes strike-slip faulting as well as crustal shortening on its margins. The analogy used to explain the fault pattern of central Asia is the deformation or extrusion of a plastic by a rigid indenter and one of the tests of the theory is that strike-slip displacement should account for much of the convergence between India and Eurasia, which is put at about 2000 km. Moreover, if the upper, brittle crust is decoupled from a ductile lower crust and upper mantle, the amount of shortening needed to lift the Tibetan plateau up to 500 m is about 100%, and the observed fold and fault displacements are far less than this.

It was a discrepancy between the observed fault displacement and that predicted by magnetic data from the oceans which led to the view that creep might be occurring at the boundary of the Iran–Arabia plates. The convergence was put at $4 \cdot 5$ cm/year averaged over the last 10 million years, which amounts to a strain rate of $0 \cdot 5 \times 10^{14}$ s^{-1} for a collision zone about 250 km wide. This is much higher than the rate obtained for twentieth-century earthquakes, but the deficit is not to be detected by strainmeter and creepmeter measurements or creep displacements on faults. The field significance of the rates thus remains a puzzle (Tchalenko, 1975). The answer may lie, as in the Tibetan example, in deformation far from the boundary. Buckling of the Indo-Australian plateau has been revealed by folds and high-angle faults, *in situ* stress determinations, and heat flow measurement (Weissel *et al.*, 1980). In the USA the prevalence of an E–W extensional field in the area between the Rockies and the Sierra Nevada is thought to indicate an upcurrent in the mantle circulation thereabouts and hence active driving of the North American plate by the circulation (Gough, 1984).

In short, the contribution to geophysics made by the analysis of recent crustal movements may turn out to be more valuable away from self-evident plate margins and within the zones currently viewed as unruffled by the vagaries of drift.

Chapter Eight

SEISMOLOGY AND DEFORMATION

*Hours before dawn we were woken by the quake . . . Then the long
pause and then the bigger shake.*

William Empson, Aubade

The seismologist is hampered by the brevity and patchy nature of
his records. The first systematic attempt to apply physical principles
to earthquake effects—that by Robert Mallet in Italy—dates only from
1857. Seismographs that we would regard as effective were not developed
until 1880. By 1960 there were about 700 stations in regular operation
throughout the globe, but the global coverage was predictably uneven,
with 130 stations in North America, 15 in South America, 120 in Japan
and 18 in Africa. Some of the stations were very recent creations; a
few, such as the seismographs operated by Antarctic and other expeditions,
were transitory. Different types of instrument were used, whence
the creation of the Worldwide Standardized Seismograph Network
(WWSSN), which by 1969 included some 120 stations in 60 countries
(Bolt, 1978).

However well equipped and run, the seismic station will record only
ground motion of one sort or another. Yet, as Richter (1958) amply
demonstrates, seismology is a dual science in which physics and geology
are interdependent (cf. Sieh, 1981). The link is most obvious in the
elucidation of ambiguous fault-plane solutions, but where the study of recent
crustal movements truly comes into its own is in the construction of seismic
chronologies, the identification of seismic gaps and the study of recurrence
intervals. Whence comes the observation that 'the geomorphology of an
area of raised beaches takes on a new light for those interested in
paleoseismicity' (Isacks *et al.*, 1968, pp. 102–103).

165

SEISMICITY AND STRUCTURE

The relationship between earthquake activity and geological structure is evidently of fundamental interest to the seismologist. It is implicit in much of his routine work. Thus earthquakes will be interpreted in the context of the regional grain whether or not they are accompanied by ground breaks. In the 1980 Italian earthquake the fault plane was selected by reference to the geological setting as well as the alignment of the aftershocks, the elongation of the damage area and the pattern of historical epicentres (Deschamps and King, 1983). Again, of three earthquakes that occurred in 1979 near the Doruneh Fault in eastern Iran, none produced ground breaks. The event of 9 December had an epicentre on the fault trace, which was visible on the ground as well as from the air. There was geomorphological evidence of sinistral strike-slip movement on the western part of the fault. The first motion data were compatible both with left-lateral motion and with a thrust mechanism. The former has the virtue of yielding a ENE slip vector, which is consistent with the results obtained at many other locations in the region (Berberian, 1976; Jackson and McKenzie, 1984, p. 231). Compare the Buyin Zara earthquake of 1 September 1962 in northern Iran, where a fault 85 km long displayed thrust faulting with a southerly dip and a sinistral strike-slip component, in good agreement with the fault-plane solution (McKenzie, 1972, p. 169) and indicative of a NE slip vector.

On a more complex level, though often still chiefly with the aim of refining or checking fault-plane solutions, is the procedure known as wave-form modelling whereby synthetic seismic traces are computed on the basis of the postulated mechanism and commonly on the assumption that the energy source is a point (Fig. 66). Comparison of the computed with the observed waveform allows depth to be derived by successive approximations to within ± 15 km and sometimes better.

Comparison of the amplitude of the two sets of waves also gives a measure of the seismic moment, that is to say the amount of energy released by the event. But the usual method is to multiply the area A of the fault displaced by the mean slip $\bar{u}$ it underwent and a measure of the rigidity μ (shear modulus) of the local rock, viz.

$$M_0 = \mu A \bar{u}.$$

Displacement can be estimated with conviction if there is surface faulting: A is often equated with the area of the aftershock zone (Molnar and Chen, 1982).* An array of portable seismographs is as essential of this purpose

*Whereas energy is usually calculated from M, Kanamori (1978) has proposed a new value of M (M_w) from the energy released in an earthquake. Here again the calculation takes average offset and the area of the fault plane into account.

as it was in improving the accuracy of the locations. The layout of the temporary stations, given uniform accessibility, will usually be governed by the pattern of ground deformation: the aim is usually to encircle the fault outcrop but at a distance that will permit the location of sufficient events for delineating the fault plane with confidence. Having to shift the stations because they were too near or too far means the loss of valuable record. But if the fault does not break the surface, reliance may once again be placed on the potentially misleading pattern of damage.

At El Asnam, in 1980, an aftershock array set up 10 days after the main event surrounded an area of 350 km^2 in which the main surface breaks occurred (Yielding *et al.*, 1981). Its results (Fig. 94) made it possible to relocate earlier events to an estimated ±5 km and placed the corrected epicentre of the main shock at the centre of the damage zone and not 70 km further north as suggested by the first teleseismic readings. The aftershocks themselves were found to be aligned with the 30 km of surface faulting. Towards the end of the 5-week recording period the network was extended

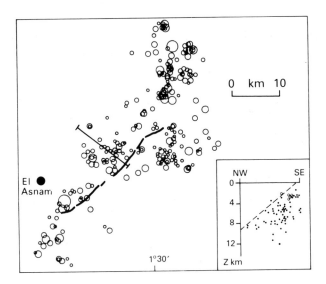

Figure 94. Aftershocks located during 1 month following the 1980 earthquake at El Asnam (Algeria). The inset shows the distribution of the better located events at depth along the section indicated by a bar on the map. Note that the section crosses a clear surface break (bold line) and that the aftershocks lie beneath the fault plane. According to King and Yielding (1984), on whose paper the figure is based, aftershocks are to be expected where the main event increases rather than decreases stress, that is to say not on the fault plane.

towards the northeast because there were indications that rupture had propagated in that direction and also to investigate the smaller displacement undergone by the most northern of the three segments into which the fault could be subdivided. By then the team were armed with a working hypothesis to account for the distribution of tensional and compressional features whereby thrusting to the southeast was accompanied by graben formation parallel to the fault.

Relocation of the main shock and of the large aftershocks, with estimated errors of 6 km or less for location and for the most part less than 10 km for depth, showed that the main shock occurred at the southwest end of the fault, propagated 12 km to the northeast where a second rupture of similar moment occurred, and continued a further 12 km northeast. The fault orientation permitted selection of the nodal plane in the solution for the main shock corresponding to a slip vector on a plane dipping to the northwest (145°). Synthetic seismograms give double wave forms similar to those recorded either by oblique slip or two successive ruptures 12 km and about four seconds apart. In other words, the inference of dual rupture is based on waveform analysis, shock location and surface deformation.

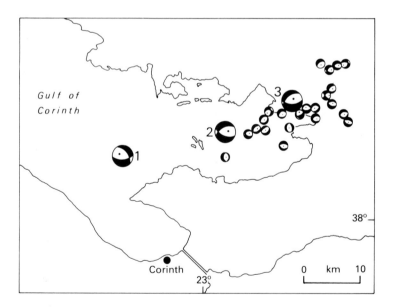

Figure 95. Fault-plane solutions for the three major events and some of the aftershocks in the Gulf of Corinth earthquakes of 1981. Note that they all indicate normal faulting, although with variable strike-slip components, and that only two indicate E-W extension. After King et al. (1985).

Calculations of the seismic moment likewise combined geomorphological data (here the length of thrusting observed at the surface, about 27 km) and the fault-plane solution (a dip of 45-60°) with aftershock data (depth of about 15 km) and waveform analysis (M_0 = total moment in the order of $2 \cdot 5 \times 10^{25}$ dyn cm).

Any attempt to relate seismic data to the regional structure must also ensure that concealed faults are included. The problems of burial by later sediments or overhanging material are familiar to field geologists, although insufficient attention has been paid to the selective survival of features (such as the grabens at El Asnam) which are in bedrock where others (here the products of reverse faulting) are in unconsolidated alluvium and may become obscured in the space of a few days. The field evidence obtained following the Corinth earthquakes of 1981 confirmed the long-standing conviction that the topography was largely the product of normal faulting and, together with the teleseismic and local seismic data (Fig. 95), indicated that the first two shocks occurred on major north-dipping faults whereas the third affected an antithetic fault dipping south (Jackson *et al.*, 1982*a*). Shifts of the coast associated with the earthquake were consistent with the proposed structure provided one took into consideration uplift of the footwall.

In seeking to trace the fault system away from the Gulf of Corinth, the ideas formulated at El Asnam were used to identify subsurface faults from surface folds and from tilted terraces and other topographic features (Vita-Finzi and King, 1985). The resulting structural pattern led to the proposition that primary faulting did not reach the surface where subsurface faults intersect or bend. It is here that we observe folding and secondary faults. Recognition of this pattern and dating of deformed or displaced marine features on the coast where possible by archaeological methods and ^{14}C assay helped to identify two major fault zones which have been seismically active at successive times during the last 40 000 years (Figs 96, 97 and 98).

RECURRENCE INTERVALS AND SEISMIC GAPS

The northern of the two fault systems in the Gulf of Corinth could be explained by a sequence of earthquakes comparable with that of 1981 and taking place every 300 years or so, for this would suffice to lift marine terraces 30 000 years old to their present maximum elevation of about 220 m. The topography of the 200 m high Pondeba ridge above the El Asnam reverse fault (King and Vita-Finzi, 1981) represents 70 uplift

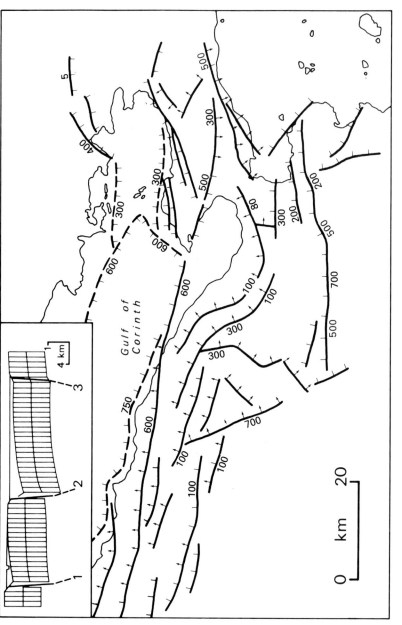

Figure 96. Faults (teeth) and downwarps (arrows) at the eastern end of the Gulf of Corinth, mainly derived from morphological evidence. The numbers denote the estimated throw. After Vita-Finzi and King (1985). Inset: Main features of the model for the Gulf described by Jackson et al. (1982a). The section illustrated runs roughly NNE–SSW (left-right) through the Isthmus of Corinth (between faults 1 and 2). Note uplift of footwall at the major faults.

Figure 97. Shallow-water marine deposits west of Corinth, at an elevation of 25 m and dated by ^{14}C to about 30 000 years BP, resting unconformably on Tertiary gravel-capped beds (just above hammer).

episodes similar to that of 10 October 1980. At first the recurrence interval was put at about 500 years; ^{14}C dating (see above p. 55) suggests it could be less than 330 years. The temptation is strong to use such results to calculate the corresponding time required for primarily tectonic landforms to be produced. If the seismic tempo has persisted throughout the late Quaternary, the eastern Gulf of Corinth could thus have formed in 180 000 years and the Pondeba ridge in a mere 230 000 years or so.

a

b

Figure 98. (a) *Chapel at Strava, near Corinth, shortly after the 1981 earthquakes. This part of the coast subsided by 1·5 m; other parts subsided by smaller amounts or were uplifted by up to 20 cm.* (b) *Partially submerged ruins of* Kenchreai, *near the eastern end of the Corinth Canal. Some 2 m of submergence, most of it post-Augustan, are indicated.*

In areas with a long historical record it may be possible to acquire the requisite number of events without recourse to field data. In northern Syria, the accounts of Greek and Roman historians from AD 52 to 588, Arabic texts of the seventh to eighteenth centuries, nineteenth-century reports and instrumental observations (Poirier *et al.*, 1980) indicate two main classes of earthquake, one with intensities on the modified Mercalli scale of VII–X (see Table 3) and the other with intensities equal to or greater than X. Where reliance is placed on geological evidence, the study is perforce generally limited to the larger events, as they are likely to leave a distinctive imprint in the geological and geomorphological record and their effects can be compared with historical equivalents. Yet for some years there has been a suspicion that particular faults or stretches of fault are visited by earthquakes which vary little in magnitude. The concept of 'characteristic earthquake' is an attractive one as regards the confidence with which the calculation is made and also its usefulness to the potential clientele. There is evidence in its favour to be found in the San Andreas and Wasatch Fault zones.

The sections on the San Andreas Fault at Pallett Creek that were investigated by Sieh contained datable wood, charcoal and peat. What is more, buried scarps and sand volcanoes could be used to identify individual episodes of movement. This kind of evidence is often seen as an alternative to historical sources. In the present example it is superior: although the occurrence and magnitude of the earthquakes has to be inferred from indirect evidence, rather than eyewitness accounts, there is no doubt about their association with fault movement (Sieh, 1978*a*, 1984).

Sieh excavated several trenches into a terrace composed of marsh deposits at a point where rapid sedimentation ensured that successive events could be distinguished in the section. The sequence included several phases of sandblow (or sand volcano) eruption, the product of seismic shaking acting on a liquified subsurface layer. Some of the faults could be seen to terminate within the section. Small scarps were also displayed in section. Most of the ^{14}C dates were on peat and the rest were on wood and charcoal. Sedimentation rates calculated between successive dated layers were used for linking the dates to the seismic events inferred from the breaks and sandblows.

Initially the evidence led Sieh to conclude, among other things, that between the sixth century AD and the 1857 earthquake ($M = 8 \cdot 25$), there had been at this locality at least six earthquakes similar in their effects to the 1857 event, and that the intervals between them had ranged between 59 ± 9 and 275 ± 68 years (Fig. 99). Further work indicated 12 earthquakes between AD 260 and 1857 with an average recurrence interval of about 145 years of which five had lateral offsets comparable to those produced

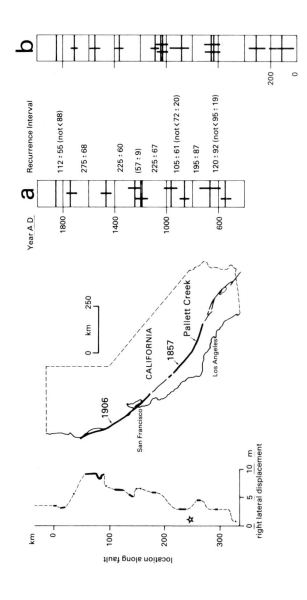

Figure 99. Left: Right-lateral offsets associated with the 1857 earthquake on the San Andreas Fault. Note variations in displacement at different points along the fault. Dotted lines are interpolated; the star marks the position of Pallett Creek. Centre: Fault ruptures of the 1857 and 1906 earthquakes. Right: Dated events at Pallett Creek. Bold horizontal lines indicate events; the vertical bars denote 1 standard deviation. (a) After Sieh (1978a); (b) after data in Sieh (1984). Note additions and revisions occasioned by fresh excavations and new radiocarbon dates.

in 1857. It may well be that events of this magnitude, with displacements of several metres, have dominated the Holocene history of the south-central part of the San Andreas Fault. Studies of similar quality elsewhere along the fault are now required to see how other parts of the fault compare.

The Wasatch Fault has not ruptured during historic times but its Holocene development has been traced by trenching scarps at various points along the fault zone and identifying successive faulting events by reference to buried colluvial wedges and graben deposits dated by [14]C. In due course six fault segments were recognized, ranging in length from about 30 to 70 km and differing in their seismic history for the last 13 500 years. The proposed sections, which were based on geomorphological observations and inferences about earthquake disturbance in the sections, showed some

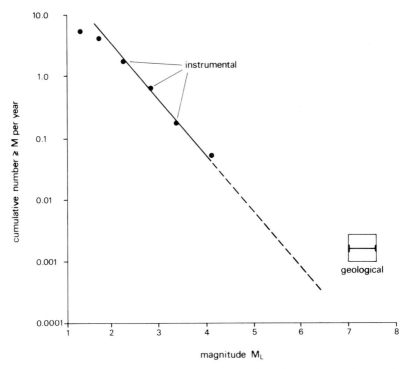

Figure 100. Cumulative frequency-magnitude plot of instrumental seismicity for a fault. The broken line is extrapolated from the instrumental record. The box represents the recurrence interval and magnitudes (here 7-7·5) indicated by the geological data. Note that the box does not lie on the line: the slope (or b value) of the graph derived from the lower magnitudes tends to underestimate the frequency of occurrence of the large uncharacteristic events. After Schwartz and Coppersmith (1984), with permission. © The American Geophysical Union.

agreement with changes in the trend of the fault zone, structural features in the bedrock, gravity data and geodetic surveys.

All the identifiable events were large, with displacements of $1 \cdot 6$–$2 \cdot 6$ m. The recurrence intervals for the four segments which bore traces of two or more events were 2000, 2400–3000, 1700–2600 and 1700–2700 years. Grouping all the data gave an interval of 400–666 years for the zone as a whole. The magnitude of the events was calculated by deriving the moment magnitude M_0 from assumed segment lengths of 35–70 km, a downdip fault width of 7–12 km and an average displacement of 2 m, and using the moment magnitude relationship $(M_0 = 1 \cdot 1 M_L + 18 \cdot 4)$ previously proposed for the Utah area.

The result confirmed the lack of evidence for small events as it was confined to magnitudes of 7–7·5. Yet a recurrence plot of instrumental data for the period 1850–1984 shows no earthquake of magnitude greater than about 5·7 within a belt running 50 km west and 25 km east of the fault, and extrapolating the recurrence curve obtained for small events (Fig. 100) might lead one to expect moderate-sized earthquakes (Schwartz and Coppersmith, 1984). The geological data, in short, are indispensable to a realistic assessment of seismicity, especially when the period of record happens to fall during a lull in the incidence of the larger events.

Besides illuminating the relationship between structure and seismicity in a general way, the recurrence record has prompted the search for mechanisms to explain periodicities in the sequence. Although most of the correlations have later been dismissed as false, the supposed link between seismicity and earth tides continues to attract attention (Kilston and Knopoff, 1983). In southern California 33 large, shallow, strike-slip earthquakes dating from 1933–80 and with magnitudes of 5·3 or greater, together with four older historical earthquakes, were found to have statistically significant 12-hourly, lunar-fortnightly, and 18·6-year periodicities. As the sample was confined to earthquakes on faults trending northwest, the correlation could be explained by the stresses induced by daily and semi-daily tides and oceanic tides: for under such extensional E-W forces the normal stress on the faults would be reduced and the shear stress conducive to faulting increased. But nothing amounting to a triggering effect could be imputed to earth tides.

The gaps revealed by historical accounts have come to be recognized in recent years as valuable clues to regional seismicity. A gap is usually taken to mean a segment of a fault or fault zone which is unexpectedly inactive. The term is conventionally confined to large events because it is these that preoccupy the engineer and also because the smaller events may be difficult to locate or are likely to have gone unreported in the pre-instrumental record. For example, the Makran coast of Iran is considered by some to be a possible gap when its instrumental seismicity is compared with that of its extension in Pakistan (Page *et al.*, 1978). The case is reinforced by making a further comparison with the marked tectonic activity reflected in the late Quaternary record of the Makran as a whole.

The Alpine Fault of New Zealand, which extends over 500 km, has seen no major earthquake during the last 150 years in its central part between 42° and 44° S. Geological evidence confirms that the charge of 'gap' is justified, as it indicates that the fault experiences earthquakes with a magnitude of about 8 and giving rise to displacements of up to 9 m every 500 years or thereabouts (Adams, 1980). The calculation was made possible by the presence of deformed river terraces (Fig. 101) whose existence was ascribed to aggradation in response to multiple landslides triggered by earthquakes: ^{14}C dating of wood from the terraces thus yielded the recurrence interval. The size of the earthquakes was derived from the fault scarps that offset the dated terraces along the fault: magnitude was then derived from the seismic moment indicated by the geometry of the fault zone, and checked against historical data for earthquakes elsewhere on South Island which had produced ground deformation.

Not all the movement on the Alpine Fault has been seismic: warping of an alluvial surface 8000-12 000 years old, and drag deformation of schists, appears to indicate aseismic movement on the fault. The question of fault creep is one which often exercises the seismologist. Is the fault truly inactive, or is it moving imperceptibly and continuously rather than in jerks? The answer (which presents special problems where the fault is concealed or beneath the sea) could determine the physical significance of the gap.

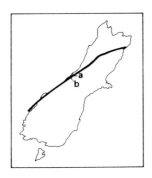
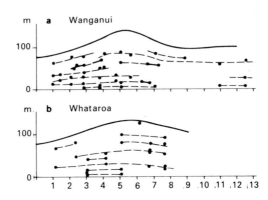

Figure 101. Profiles of two rivers normal to the Alpine Fault of South Island, New Zealand, showing aggradation surface (unbroken) *and terrace remnants (dots) above the river. After Adams (1980) with permission of The Geological Society of America.*

Although seismic gaps could be identified and discussed before the advent of plate tectonics, if only in the sense that these were areas whose seismic quiescence contrasted with the agitation that prevailed in the rest of the region, sea-floor spreading and its corollaries show how the active and inactive zones could be linked and provide a basis for calculating the strain at the various boundaries. In one attempt to match plate motion with seismicity (Rikitake, 1976) the working hypothesis was that, in subduction zones, 'great' earthquakes — that is earthquakes of about magnitude 8 — arise when continental lithosphere is dragged by an underthrusting oceanic plate until it ruptures. The recurrence interval for different parts of the Pacific subduction margins was found to range from 27 to 117 years, but the expected relationship between the interval and the local rate of plate convergence did not emerge. For example, the Aleutian–Alaska zone had a recurrence interval of $27 \cdot 2 \pm 8 \cdot 9$ years; relative plate velocity here is put at $7 \cdot 5$ cm/year. The recurrence interval for southern South America is $100 \pm 22 \cdot 5$ years and relative plate velocity is higher than in the Aleutian–Alaska zone $(8 \cdot 0)$ where one would expect the reverse. The conclusion has to be that local conditions, such as the interface geometry and the physical properties of the plates, not unexpectedly complicate matters.

At any event the uncertainty that surrounds both sides of the equation is considerable: quite apart from the likelihood that the driving force fluctuates from the accepted average rate, seismic history for many parts of the Pacific coast is very vague. Yet the approach is clearly a foretaste of the explanatory models that increasingly will supplant statistical or geometrical assessments of gaps and that will link seismic records with the accumulating data of ground deformation. It could turn out that the concept of gaps is misleading, resting as it does on the apparent lack of overlap between the sources of adjacent earthquakes (Kisslinger, 1978). Improvements in the techniques for locating earthquake sources combined with detailed mapping of ground deformation will in due course show whether large shocks are truly territorial. Needless to say the geological evidence takes over where the historical records are too short for the recurrence intervals, notably in continental areas. Figures of hundreds to thousands of years are cited for recurrence intervals in Nevada and Utah (Scholz *et al.*, 1973; Simpson, 1980).

SEISMIC CYCLES

In 1969 Soviet seismologists working in central Asia noted that the ratio between the velocity of P (compressional) and S (shear) waves changed before a series of earthquakes. A similar effect was later noted in the Adirondack area of New York State and 3½ years before the San Fernando earthquake of February 1971. The explanation that came to be favoured ascribed the effect to dilatancy, the increase in the volume of a rock when it is stressed but before it fails. The expansion is inelastic. The resulting pore space and crack development render the rock undersaturated but pore pressure in due course rises again as the stress continues to increase until the earthquake takes place. Thus dilatancy probably delays the earthquake while the fluid pressure on the fault is reduced. It is clear that seismic velocities will be affected by such a sequence. It is also obvious that other physical changes are likely to become measurable. This is one great attraction of the dilatancy model.

Some of the precursors in question are governed by the hydrological effects of dilatancy, such as changes in electrical resistivity and the emission of radon and other short-lived isotopes. The premonitory seismic pattern mentioned earlier could also be a product of fluctuations in pore pressure. Above all, dilatancy implies temporary uplift to an extent that should be detectable by levelling or the use of tiltmeters.

A classic instance of geodetic premonition is the 1964 Niigata earthquake in Japan ($M = 7 \cdot 5$). Geodetic resurveys showed steady vertical movements from 1898 to 1955, presumably the period of strain accumulation, followed by rapid uplift in 1958 by almost 5 cm, which may well indicate dilatation. There ensued a relatively stable period of 5 years, when pore fluids were supposedly flowing into the cracks, and then the earthquake. That uplift was absolute is shown by a corresponding fall in sea level at Nezugaseki, about 100 km away (Fig. 102). At distances greater than this the uplift was not observed. Anomalous changes in the magnetic field accompanied the sequence, a fact which supports the inference that pore water was undergoing redistribution.

The earthquake ($M_L = 6 \cdot 4$) that struck San Fernando, California, on 9 February 1971, was also preceded by vertical crustal movements. These affected a large part of the Transverse Ranges. Levelling data showed predominantly upward and generally episodic movement, with a maximum uplift of $0 \cdot 2$ m between 1960-1 and 1968-9 about 30 km to the northwest of the 1971 epicentre. The fault plane solution indicated slip on a reverse fault dipping to the northeast, and it was found reasonable to explain the episodic movements by deep-seated creep on the fault. But there was some evidence that dilatancy was implicated in the coincidence between uplift of the epicentral area in 1968-9 and the onset of an anomalous ratio in the P/S seismic velocities (Castle *et al.*, 1975; see also Thatcher, 1984). The latter view permits the San Fernando earthquake to be set in the context of the Palmdale aseismic uplift of the early 1960s. Uplift is then ascribed to stress accumulation along the San Andreas Fault and failure on nearby reverse faults is a secondary effect

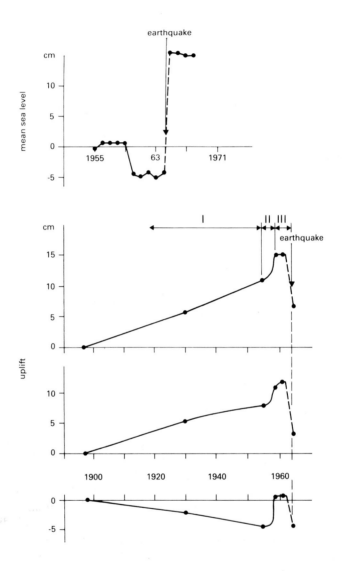

Figure 102. The top curve shows relative sea level near Niigata, Japan. The other three curves show elevation differences for three bench marks resurveyed repeatedly before the 1964 earthquake (M = 7·5). Note the correspondence with premonitory stages postulated by the dilatancy model, viz (I) elastic strain accumulation, (II) dilatancy dominant, (III) influx of water dominant. After Scholz et al. (1973) with permission. ©1973 The American Association for the Advancement of Science.

produced by increased shear stress at a time when it has locked the main (San Andreas) fault.

There are often explanations for pre-earthquake deformation besides dilatancy and deep-seated creep. A view held by some Soviet geophysicists is that, whereas the growth of cracks is involved, the diffusion of water into and out of the focal region is not required to explain the observed sequence of events: the earthquake occurs during a period of decreasing stress and any concomitant increase in seismic velocity stems from crack closure during the fall in stress. Granted that careful monitoring of electrical and seismological variables is required if the two competing views are to be evaluated, a crucial difference that could be amenable to field testing is that, whereas the dilatancy model assumes that an earthquake can occur on an existing fault and does not require large-scale fracture, in the 'dry' model the earthquake occurs when a fault is being created (Mjachkin *et al.*, 1975).

If they are to provide a useful test of a proposed seismic mechanism, the earthquake precursors need to be supplemented by a full account of the changes that accompanied and followed the earthquake. Coseismic and postseismic field studies are likely to be hampered by destruction and confusion and by the need to give priority to medical and social assistance; on the other hand, now that the epicentre or at any rate the epicentral area has revealed itself, the work can be focused on the key locations and geared to the most promising sources.

There is general agreement that shallow earthquakes generally represent the release of strain in the form of faulting. Nevertheless the deformation patterns observed after earthquakes vary widely to include aftershocks and creep, and, besides attempts to assess the influence of the local stress field and the nature of the crust, various ancillary mechanisms such as aseismic slip at depth and viscoelastic flow have been invoked to explain the diversity of modes of strain release (Kanamori, 1973). It is not always possible to distinguish between dilatancy and creep, for example. The Palmdale bulge of southern California can be explained by either (Wyss, 1977).

Three shallow strike-slip earthquakes in Japan, at Tottori, Kukui and Tango, indicated more or less instantaneous stress release and thus lend weight to the Reid model of simple elastic rebound. The Niigata earthquake, associated with dip-slip faulting, was also consistent with the release of accumulated strain energy by rupture. All four earthquakes may thus indicate fault motion confined to depths of less than 20 km where brittle behaviour prevails. In contrast, large aseismic deformation characterized the disastrous Kanto (Kwanto) earthquake of 1 September 1923 ($M = 8 \cdot 2$), and some viscoelastic behaviour is indicated. Seismic fault movement, which was a combination of dip-slip and right-lateral, amounted to 2 m, whereas geodetic data indicated about 7 m of slip. Coseismic uplift, which locally attained almost 2 m, was followed by gradual subsidence (Matuzawa, 1964).

The problem remains of accounting for the non-elastic behaviour. Unlike other large Japanese earthquakes with a large aseismic element, the Kanto record cannot be explained by activity at a depth sufficient to bring mantle behaviour into the picture: the Nankaido earthquake of 1946, for instance, is thought to stem from underthrusting of Honshu Island by the Philippine Sea plate down to a depth of 80 km. At Kanto the interaction between the Philippine Sea plate and Honshu Island takes the form of right-lateral slip. One suggestion for this kind of behaviour is that the lithosphere is anomalously thin in major fault zones. Another is that, perhaps owing to raised isotherms near tectonic plate boundaries, the lower lithosphere displays viscoelastic properties on time scales greater than that of the earthquake cycle (Cohen, 1984).

Detailed analysis of ground movements before, during and after an earthquake proved crucial to interpretation of the 1964 Alaska earthquake ($M = 8 \cdot 4 +$). The primary fault was not exposed on land and accordingly two fault models were proposed (Plafker, 1969): relative seaward thrusting

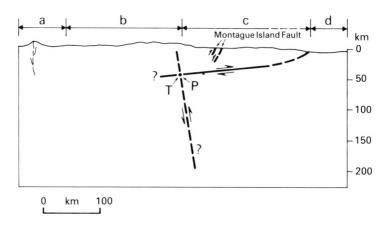

Figure 103. The 1964 Alaska earthquake. Two interpretations based on residual vertical movements and focal mechanisms. The section is approximately normal to the regional structural trend. P and T are respectively pressure and tension axes at the hypocentre of the main shock. According to one view the cause was thrusting along a fault which dips northwestwards at a low angle beneath the continental margin; according to the other, dip-slip movement on an almost vertical fault with the upthrown side near the ocean and striking roughly along the zero isobase (see also Fig. 43). (a) slight uplift, (b) major zone of subsidence (c) major zone of uplift (d) possibly subsidence. The Aleutian Trench axis lies between c and d. After Plafker (1969).

along a fault dipping northwest beneath the continental margin at a low angle, and dip-slip movement on a steep vertical fault with the seaward side upthrown (Fig. 103). The former alternative gained support from information on movements before and after the earthquake and from fault-plane analysis of the main shocks and longer aftershocks.

The horizontal displacements associated with the earthquake and the coastal·submergence that preceded it suggest that shear failure followed 930–1360 years of elastic stain accumulation. The drive may have been provided by underthrusting: the fault-plane solutions for the main shock and many of the larger aftershocks are consistent with the pattern of shearing one would expect as a consequence of this kind of drag. The earlier stages of strain accumulation could be accommodated by local movements: none of the pre-1964 earthquakes with magnitudes of 7 or more were accompanied by regional deformation.

Elastic rebound after failure would account for uplift on the coast and subsidence inland. There remain unexplained instances of uplift and an asymmetry in the volumes of uplift and subsidence, but until the offshore record is known in detail the significance of such discrepancies is not clear. The steep-fault models finds no convincing support in the geological, geodetic or seismological evidence.

Chapter Nine

FUTURE EARTH
MOVEMENTS

*The physician can bury his mistakes, but the architect can only
advise his client to plant vines.*

Frank Lloyd Wright

Seismological research has always gained impetus from the belief (not always
shared by those engaged in the work) that the results would in due course
make it possible to predict the size, location and timing of earthquakes
(Fig. 104). As interest in precursory movements continues to grow, the
study of creep and other non-seismic modes of deformation, hitherto largely
the preserve of surveyors and academic scientists, begins to acquire a similar
urgency. As evidence for such gradual movements mounts up, so do their
cumulative effects raise questions about the long-term safety or effectiveness
of nuclear installations, dams and sea walls in areas previously considered
to be secure.

Risk begets investment. Japan, battered by earthquakes and tsunamis,
maintains several institutes devoted to seismic research. China, the USSR
and the USA are understandably prominent in exploring the engineering
implications of crustal instability. Geodesists in the Low Countries have
long kept the configuration of the North Sea under scrutiny. But the map
of indifference is being redrawn by advances in knowledge and technology
as well as by the imposition of more stringent and uniform safety standards.
For example, the United Kingdom, as noted in the Preface, is now
represented in the International Association for Earthquake Engineering,
and the membership is by no means confined to designers and opponents
of nuclear power stations.

Figure 104. Precise earthquake prediction. Cartoon by Larry, reproduced with permission.

EARTHQUAKE PREDICTION

Some practitioners prefer to speak of forecasting rather than predicting earthquakes, the implication being that, as in synoptic meteorology, the complexity of the system rules out anything but general statements of the likely outcome. But the subtlety would escape climatologists, who often use prediction and forecast interchangeably (see Cressman, 1967) or at most apply prediction to individual components of the weather and forecast to the components viewed collectively. It is more helpful (and here the analogy with meteorology is closer) to distinguish between short- and long-term predictions.

Very long-range predictions have also been called statements of earthquake potential (Wesson and Wallace, 1985). They rely on slip rates derived from the offset of geological features coupled with a model (such as Reid's) by which to explain the episodic nature of movement. Long-range predictions exploit the data of palaeoseismology as epitomized by Sieh's work on Pallett Creek. According to R. E. Wallace,

$$R_x = \frac{D}{S - C}$$

where R_x is the recurrence interval at a point on the fault at issue, D is the displacement accompanying an earthquake and related empirically to its magnitude, S is the long-term strain derived from geodetic or geological data and C is the aseismic tectonic creep rate (Wallace, 1970; Lamar *et al.*, 1973). If an entire fault system is being investigated, the following formula may be more appropriate:

$$R_t = \frac{R \times L}{L_t}$$

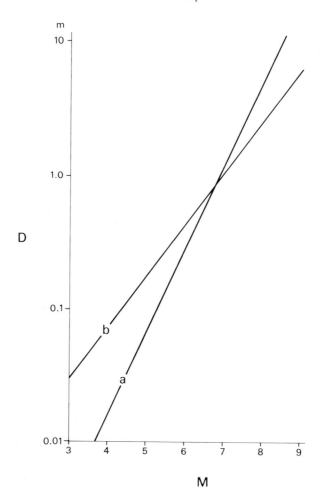

Figure 105. Maximum surface displacement on main fault (D) *plotted against earthquake magnitude* (M). (a) *strike-slip faults;* (b) *reverse faults. After Lamar* et al. *(1973).*

where R_t is the recurrence interval along the fault, L is the length of the fault rupture (which is related empirically to M_L) and L_t is the total length of fault for which the interval is sought. The results obtained for faults at individual points and for the faults as a whole may show marked differences. Fig. 105 illustrates the use of graphs for deriving fault rupture from M_L. For example, for an $M = 7$ earthquake the Garlock (strike-slip) Fault, with an average slip rate of $0 \cdot 8$ cm/year, gives a recurrence interval of about 200 years for a point on the fault and 30–90 years over a length of fault. Not surprisingly the differences are smaller or nil with reverse faults. For the White Wolf Fault at $M = 6$ event would thus give an interval of 1000 years at a point and 200–900 years for a length of fault (Lamar *et al.*, 1973).

The prediction may be confined to the likelihood that an earthquake of a particular magnitude will occur. In Syria, recurrence intervals were calculated from historical records for earthquakes with intensities of VI to less than X and for those with intensities of X or more. Seismic moment was derived from the recurrence interval and in turn used to calculate the magnitude of the two sets of earthquakes, namely *c*. $7 \cdot 6$ with a recurrence interval of 341 ± 62 years, and 6–7 with a recurrence interval of 71 ± 51. As the last large earthquake had occurred in 1822 the next one was not foreseen for another century at least, but as the last earthquake of $M = 6$–7 dated from 1872, a fresh one seemed likely to come soon (Poirier *et al.*, 1980). In the central and eastern Hellenic Arc the rate of seismicity for events with magnitudes over $4 \cdot 9$ has been fairly constant since 1950. In the west, however, the rate has fallen by about 20% since 1962. If the arc is to witness any large earthquakes during the present decade, this is where they are expected to strike (Wyss and Baer, 1981).

The characteristic earthquake model, it will be recalled, suggests that recurrence relationships calculated for individual faults on the basis of slip rate and a constant value of b (see Fig. 105) can underestimate the frequency of large events (Schwartz and Coppersmith, 1984; see above, Chapter 8). Further enlightenment has come from the concept of 'asperity', the term used for segments of a fault along which stresses accumulate. The location of asperities is derived from the distribution and type of small earthquakes and from the geometry of the fault, and it is hardly surprising that most of the work on the subject has been carried out on strike-slip faults such as the San Andreas. The successful prediction 4 years in advance of an earthquake of magnitude $6 \cdot 1$ near Morgan Hill, in California, rested on the surmise that a shock of the same magnitude in 1911 was indicative of where fresh movement was likely to recur and, by virtue of the fault geometry, that it would be of comparable size (Kerr, 1984).

As regards short-term prediction, hopes rest largely on the premonitory movements outlined in Chapter 8 combined with related geophysical or geochemical effects. In China monitoring for the purpose embraces changes in fault creep, tilt (regional as well as local), resistivity, shallow earth strains, water table level and spring flow, radon content in the groundwater, the velocity of seismic waves, the patterns of microseismicity and of foreshocks, and gravity. The 1976 Tangshan earthquake (magnitude 7·8) was preceded by changes in creep rate, gravity, strain, resistivity and well levels. But the creep phenomena, which were not noticed until after the earthquake, occurred on the Babaashan Fault 180-200 km away from the epicentre, and the resistivity data were obtained 80 km east of Tangshan. There were no foreshocks. At both Lungling and Yenyen, however, foreshocks led seismologists to the epicentral area weeks ahead of the main shock, and other premonitory clues were consequently recognized (Lommitz and Lommitz, 1978; Sacks *et al.*, 1978).

The analysis of premonitory deformation, coupled with observations on coseismic and postseismic movements, contributes to earthquake prediction in an additional, roundabout way: by improving palaeoseismological analysis. South of Tokyo, four marine terraces rising to 25 m date from the last 8000 years. They are interpreted as the products of three successive earthquakes similar in size to the 1923 event during which the lowest terrace was uplifted; to judge from this event, they were followed by subsidence at 1-2 mm/year, and the uplift they now indicate must be corrected accordingly. Moreover, although the oldest terrace is at roughly the same height in the Oiso region, southwest of Tokyo, as it is further east, on the Boso peninsula, the combined coseismic uplift in the 1703 and 1923 earthquakes in the former region was 2 m whereas it was 5-6 m in the Boso area. The discrepancy is seen as a possible indication that the Oiso area is a candidate for a future large earthquake (Matsuda *et al.*, 1978).

Of course, prediction would need to be consistently accurate not to prove ineffective or even counterproductive. The 1975 Haicheng earthquake ($M = 7·3$) was preceded by sufficient premonitory phenomena for an urgent warning to be issued to the local population a few hours before the event and, despite extensive destruction, casualties were thus kept very low (Marshall, 1983). Yet there have been many false alarms, and no warning at all was given of the Tangshan earthquake of 1976, during which about 650 000 people were killed.

SEISMIC ZONING

In any case bald prediction is only the first step towards guarding against its effects. The impact of an earthquake or other kind of displacement depends, among other things, on its character and its size, its distance from the place under consideration, local conditions of terrain and soil, and both economic and social factors (Fig. 106). For example, the 1964 Alaska earthquake released twice as much energy as the 1906 San Francisco earthquake but whereas 700 people died in California, the low population density in Alaska kept the death toll down to 114. In California much of the damage was caused by fire, in Alaska shaking and compaction of sediments were more destructive. The Lisbon earthquake of 1755, immortalized in *Candide*, killed 30 000, many of whom were in church at the time. A single landslide during the 1970 Peru earthquake buried two towns with the death of some 20 000 people. At Niigata (1964) buildings fell over when the soil underwent consolidation and liquefaction. Lyell (1837) tells how the town of Tomboro, on the island of Sumbawa near Java, was submerged — rather than torn asunder — by an earthquake in 1815. Though spectacular and the subject of many picturesque anecdotes, seismic faulting is rarely the major source of damage during an earthquake. The surface break produced during the Californian earthquake of 1906 caused damage only to a few houses and stretches of pipeline (Richter, 1958). At El Asnam a train which was crossing the main fault during the main shock was derailed when the line was buckled by the new scarp that arose over the reversed fault, but this caused none of the 3000 deaths credited to the earthquake sequence.

The aim of seismic zoning is to subdivide a region or country into areas differing in their proneness to damaging earthquakes. The criteria used for delimiting the zones vary greatly between countries and in addition reflect advances in seismology. In the USSR the first attempt was made in 1937. A new map was published in 1957. Following several years of study a third map was published in 1968 (Medvedev, 1976). As adequate information on earthquake frequency was not available for much of the country, the map had to be confined to 'seismic activity', that is to say the distribution of earthquakes of various magnitudes. Different schemes of classification had to be harmonized so far as possible and, in view of the unevenness of the station network, gaps were filled by reference to frequency distributions.

The crucial item of data was the average recurrence period of earthquakes capable of causing earthquakes of intensities VII or VIII at the Earth's surface and with focal depths of 5-20 km. It was felt that this information was of practical interest and also sufficient for estimating the frequency of recurrence of larger earthquakes (Bune, 1976). The Soviet intensity scale permits the use of either instrumental data or ground deformation and building damage. It is equivalent in result to the Modified Mercalli scale used in Europe since 1931. Geological criteria, including information on recent movements, were then used to define zones of equal seismic

- NO KEN PRET NA LUSIM GUTPELA TINGTING. Yu no ken bagarap sapos i no gat wanpela samting i pundaun antap long yu.

- SAPOS YU STAP INSAIT LONG HAUS, YU NO MAS GO AUSAIT. Stap na hait long hap yu makim pinis.

- NO KEN YUSIM PAIA olsem long kukim kaikai, na lam kerasin, na kandel. No ken slekim masis stret bihain long guria. Mekim dai olgeta paia.

- SAPOS YU STAP AUSAIT, no ken sanap klostu long ol haus, o bikpela diwai na ol pawa lain.

- NO KEN RAN INSAIT O KLOSTU LONG OL HAUS. Birua i save kamap planti taim long ol samting i save pundaun_ausait long dua na klostu long ol ausait banis bilong ol haus.

- SAPOS YU STAP INSAIT LONG KAR I RAN, yu mas stapim kwiktaim. Tasol yu mas sindaun i stap insait long kar inap guria i pinis.

OL SAMTING I GUT LONG MEKIM BIHAIN LONG GURIA:

- Go lukluk sapos i gat paia.

- Lukluk sapos pawa i wok na wara i ran o nogat. Lukaut long ol pawa lain i pundaun long graun. No ken holim ol.

- Sapos pawa i no ran gut, mekim dai swis long nambawan bokis bilong pawa.

- Sapos ol paip bilong wara i bruk, pasim nambawan kok. Pasim ol bruk long ol tang wara.

- Luksave sapos ol lata i go antap long haus i bagarap na sapos ol i pas strong yet long hap bilong ol.

- No ken yusim telepon inap yu laik mekim wanpela ripot kwik.

- No ken wokabaut na lukluk long ol samting nabaut.

- No ken go klostu long ol haus i kisim bikpela bagarap o ol hap graun i bin lus na pundaun nating. Bihain long guria ol inap long pundaun yet.

BIRUA NOGUT TRU I KEN KAM LONG HAP BILONG SOLWARA BIHAIN LONG GURIA

Sapos guria i pinis yu mas lusim nambis na go kwiktaim long ples i antap na yu mas stap long hap inap long wanpela aua. Sapos solwara i surik na rip i stap ausait YU NO KEN GO WOKABAUT LONG OL RIP. Bikpela haiwara tru baimbai i mas kamap kwiktaim.

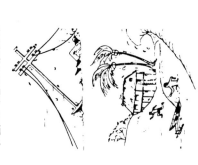

- DON'T PANIC. The motion is frightening. You are safe unless something falls on you.
- IF YOU ARE INDOORS, STAY INDOORS. Take cover in a previously chosen safe place.
- DON'T USE OPEN FLAMES such as kero lamps, candles, or matches, either during or after the tremor.
- IF YOU ARE OUTSIDE, move away from buildings, large trees and electricity wires. Once in the open, stay there until the shaking stops.
- DON'T RUN THROUGH OR NEAR BUILDINGS. The greatest danger from falling debris is just outside doorways and close to outer walls.
- IF YOU ARE IN A MOVING CAR, stop as quickly as safety permits, but stay in the vehicle. It is good place to stay until the shaking stops.

WHAT TO DO AFTER EARTHQUAKE:

- Check for fire.
- Check your electricity and water supply. Earth movement may have cracked water and electrical conduits. Look for fallen power lines.
- If electrical wiring is damaged, switch off at the main meter box.
- If water mains are damaged, shut off the supply at the main valve. Plug leaking water tanks.
- Check to see whether stairways are safely in place.
- Stay off the telephone except to report an emergency.
- Don't go sight-seeing.
- Stay out of severely damaged buildings or land slip areas. Aftershocks can shake them down.

AN EARTHQUAKE'S WORST KILLER MAY COME FROM THE SEA

In a low-lying coastal area, an earthquake may be your only warning that a tidal wave is about to strike. As soon as the shaking stops, start for high ground and stay there for one hour. If the sea goes out DO NOT walk onto the reefs because a tidal wave is sure to occur.

Figure 106. Advice to coastal dwellers in pidgin. Note in particular the warning about seismic sea waves (bikpela haiwara).

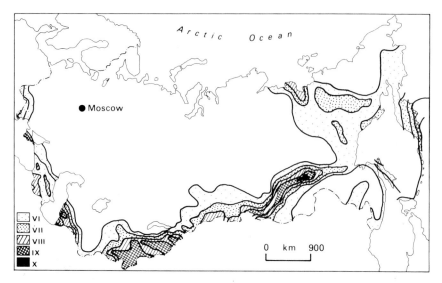

Figure 107. Seismic zoning map of the USSR prepared in 1967 and based on the All-Union State Standard (GOST) intensity scale. After Medvedev (1976), with permission from Keter Publishing House, Jerusalem Ltd. (also Fig. 108a).

risk. The final map (Fig. 107) shows zones in which, under average ground conditions, shaking may occur during earthquakes of intensity VI or more.

The vagueness of this approach is a measure of the problems facing the creators of seismic risk maps. In the USA one such map drawn in 1969 identified four zones on the basis of the largest earthquakes they had experienced. The earthquakes were classified using the MM scale and 'evidence of strain release' and geological features thought to be associated with seismicity helped in the delineation of the zones (Howell, 1973), but there are obvious drawbacks in the reliance on what must be an imperfect record. An earlier proposal by Richter (1958; see Medvedev, 1965) used MM intensities 6-9 combined with some geological data [Fig. 108(a).].

A map drawn up in 1976-7 finally grasped the nettle and considered risk directly rather than by implication: it showed the effective maximum or peak acceleration to be expected at odds of 1 in 10 during a 50-year period (Fig. 108). In other words, the chances are 9 in 10 that a particular value at a point on the map will not be exceeded in 50 years. The map takes into account historical data, major structural features and the distance—decay effect (or attenuation) displayed by intensity away from the fault (Bolt, 1978, pp. 175-176). An alternative, used in Chile, is to show the probability, as a percentage, that a particular value of acceleration, such as $0 \cdot 1g$, will occur during a specified number of years (Lommitz,

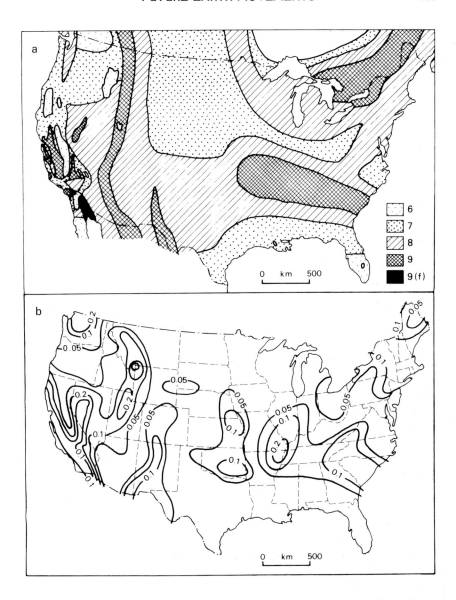

Figure 108. (a) Seismic zoning map of the USA broadly comparable with that in Fig. 107, as it is based mainly on the Modified Mercalli scale. 9f denotes regions where earthquakes of intensity 9 are frequent. After Richter (1958), reproduced in Medvedev (1965). (b) Seismic risk map of the USA drawn up in 1976-7. The contours indicate effective maximum acceleration levels expressed as fractions of gravity to be expected with p = 1 in 10 in a 50-year period. After Bolt (1978) with permission. © 1978 W. H. Freeman & Co.

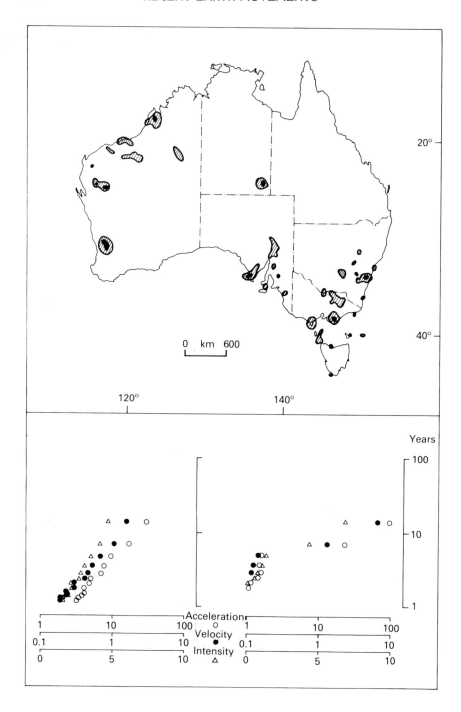

1974, p. 116). Acceleration of $0 \cdot 1g$ or more, especially if sustained over several seconds, is considered damaging to structures not designed to resist it. The greatest injury often stems from related effects, such as the flow of so-called quickclays whose strength has been drastically reduced when interparticle bonds are destroyed by vibration (Smalley, 1967). But the zonation still represents an advance on maps of earthquake magnitude.

A similar approach has been followed in Australia. Besides ground acceleration it computes velocity and intensity at the modes of a half-degree grid and plots the results against the recurrence interval on logarithmic scales (McEwin *et al.*, 1976). Extrapolated values with a 50-year return period are then plotted and contoured on maps (Fig. 109). The validity of extrapolating to 50 years from a record of only 13 years is less laughable than at first appears. Quite apart from the need to do just that when the record is short, it is often the case that earthquakes follow a Poisson distribution. Where the record for the earth as a whole is concerned the match is improved by disregarding earthquakes which represent the aftershocks of major events. Low magnitude events may need to be grouped or 'clustered' before plotting (Scheidegger, 1975). If a Poisson process is indeed apparent the probability that a certain number of earthquakes will occur during a specific period can be calculated. On the other hand, if the record departs from such a distribution, one may legitimately conclude that the activity has fluctuated. In New England, for example, the level of seismicity was apparently higher during the period 1725–1849 than during 1849–1974. It is then still possible to compute recurrence intervals by using the extreme-value method, which relies on the largest events and is thus appropriate for historical data (Shakal and Toksöz, 1977).

Microzoning is the name given to the procedure by which surface and subsurface conditions are used to assess the damage likely to be produced by seismic vibrations in different parts of the area being surveyed. It is strongly influenced by the development of building codes (Lomnitz, 1974; see also Clark and Persoage, 1970) and in its turn enters into their formulation. In the Wellington area three kinds of sediment were considered for the purposes of microzoning: basement rock, compact sediment and

---

Figure 109 (opposite). Upper: Ground acceleration in Australia (50-year return period). The hatched areas represent areas with accelerations 1–3 m/s², the black areas accelerations of $\geqslant 3$ m/s². Lower: Computer plots of acceleration (cgs), velocity (cgs) and intensity (Modified Mercalli scale) against return period in years for two sites. The one on the left shows a good correlation and thus encourages extrapolation to the 50-year period; the site on the right does not. After McEwin et al. (1967) with permission from the Director of the Bureau of Mineral Resources, Canberra.

high-porosity sediments (Grant-Taylor *et al.*, 1974). The last are likely to amplify earthquake waves more than the other two categories; some high-porosity sediments may even flow under vibration. The survey also distinguishes areas likely to fail by virtue of their steep slopes and, of course, those faults known to be active.

Because it relies heavily on informed conjecture about the likely size, depth and duration of events in the future, microzoning tends to be dismissed as of limited value. But it is of great significance to the urban planner, because it prompts the identification of high-risk zones and does so chiefly on the basis of geological evidence assessed from the viewpoint of soil mechanics. The premonitory agitation experienced by many animal species before earthquakes could owe something to ground disturbance produced by dilatant phenomena or foreshocks in relatively unstable areas.

SLIP, CREEP AND TILT

In New Zealand geologists recognize as Class I Active Faults those that have shown repeated movement over the last 5000 years or a single movement in the last 50 000 years—more succinctly, those that move sufficiently often and with displacements so large that there are official recommendations to the effect that no structure should be built across the known or presumed trace of Class I Active Faults (Grant-Taylor *et al.*, 1974; p. 12). In the USA, the Nuclear Regulatory Commission treats as 'capable' (that is to say mobile in a hazardous way) a fault which has exhibited movement at least once in the last 35 000 years or movement of a recurrent nature within the last 500 000 years (Lamar *et al.*, 1973). For the US Bureau of Reclamation the limiting age for an active fault is 35 000 years; for the State of California 11 000 years. Town Councils concerned for the safety of one- or two-floor low-density housing in California leave the decision to the geologist charged with writing the site report. Arbitrary figures of this kind provide psychological rather than rational reassurance about the future. Where movement is progressive, comfort is sought in constancy of change.

The damage suffered by the Taylor Winery near Hollister—the first indication that the fault was experiencing strike-slip movement at an appreciable rate—stemmed largely from creep. Some cracking was produced by the 1906 San Francisco earthquake and local earthquakes in 1939, 1947 and 1960, but the effects were trivial in comparison with those produced by shearing motion. As it happens the progressive damage reported by workers at the winery was initially ascribed to some kind of landslide effect but the linearity of the distortion soon showed that conventional explanations were inadequate.

The creep is not uniform. Indeed, short periods of movement lasting about 1 week are separated by weeks or months with no perceptible displacement, so that in one 371-day period 92% of the movement took place in four bursts totalling 34 days, that is in 9% of the time. But the aseismic connotations of the word creep justify its retention: for although one of the spasms began with 3 mm of dextral movement coinciding with a local earthquake, the other three were not accompanied by any seismicity in the area. Further measurements have confirmed that the accumulated slippage is remarkably uniform (Fig. 57) (Steinbrugge *et al.*, 1960; Meade, 1971).

One can think of fluids other than wine whose spillage could cause concern. Structural damage is also potentially serious. Creep on the Hayward Fault has damaged warehouses, pipelines, railway tracks and tunnels. As the phenomenon was not widely recognized until 1956 it is too late to prevent the spread of business and residential building in the San Francisco Bay area on to the creeping fault lines. But some of the local administrations now have the power to influence the design and location of new buildings. Thus in Hayward and in Fremont, also on the Hayward Fault, the city engineer can ask for proposals to be modified in the light of detailed geological reports on the sites in question (Radbruch, 1971).

The problems of progressive ground movement are just as commonplace in Pozzuoli, where the postclassical submergence and uplift reported by Babbage and Lyell continue at the present day. Since their erection in the first century AD, the columns have moved up and down over some 12 m. The *Lithodomus* borings indicate submergence by over 6 m. The site was under water in the thirteenth century. Uplift began in or soon after the sixteenth century and by 1803 the amount of water remaining was seen as a nuisance largely by virtue of the mosquitoes it bred. By 1828 the bases of the columns were 30 cm below water level. By 1878 this had increased to 65 cm, by 1913 to 1·53 and by 1954 to 2·5 m (Beloussov, 1962; Gorshkov and Yakushova, 1967). Sinking continued at about 15 mm/year until 1963 when general uplift at about 70 mm/year supervened. Between 1969 and 1972 the rise totalled 170 cm at a rate of up to 5 mm/day and affected an area of about 170 km^2 (C. Kilburn, *The Guardian* 10 November 1983; Fig. 7).

The authorities thus have to contend with movements which can change direction and speed within the space of a few years or even months. Nor are the movements wholly aseismic: on 4 October 1983, for instance, there was an earthquake of M_R 3·9, and several smaller events have punctuated the accelerated movements that began in 1969. Quite apart from the extensive damage produced by the tremors, the association of deformation with seismicity is alarming when it is recalled that the creation of Monte Nuovo, a cinder cone 120 m high, followed 2 years of increased seismic activity and a few days of uplift amounting to $6-7 \, m^3$. For there is a long-standing belief that both effects arise from pulses of magma intrusion into the roof of a chamber beneath Pozzuoli: in short, that an eruption cannot be far behind. This is of course a variant of the view expressed by Babbage and endorsed by Lyell that uplift stems largely from the instrusion of lava in turn prompted by an increase in subterranean heat. But the level of understanding is inadequate to permit confident prediction and corresponding executive decisiveness.

The uplift undergone by Fennoscandia is more predictable and appears less ominous, but its economic impact has not been trivial; it is also amenable to planning. For instance, the total land area added to Finland each century is put at about $1000 \, km^2$: 'a man in his old age may harvest where he waded as a child' and new lands are allegedly divided among landowners roughly every 50 years (Charlesworth, 1957, p. 1328). With a maximum uplift rate of about 1 m/100 years at the head of the Gulf of Bothnia, one may expect some impact on the navigability of inshore waters. Certain harbours such as that at Vaasa have had to move and others still in use in the Middle Ages are now too far inland to be serviceable.

One can try to calculate the likely extent of submergence on coasts which are undergoing tilting or warping. On the Atlantic coast of Canada the agreement between the archaeological and tide-gauge evidence (p. 102) inspires some confidence. Where the trend is still in dispute, as on several Pacific islands (Bloom, 1980), the engineer must presumably guard against every contingency. It is perhaps surprising to find that the coast of southeast England falls into this category.

For many years the accepted view has been that, in sympathy with other parts of the North Sea, the Channel coast of England subsided during the Holocene at a time when the north and west of the country was undergoing uplift. Attempts to quantify this belief have relied on radiocarbon dating of intertidal peats, the position of archaeological remains and tidal records (Churchill, 1965; Smalley, 1967; Rossiter, 1972) (see Chapter 5). The struggle against the sea by coastal communities seemed destined to grow grimmer and devices such as the Thames Barrage to be cursed with an uncertain future. But the evidence for recent subsidence is itself by no means secure. The geodetic sources are inconclusive (Kelsey, 1972; Shennan, 1983), and there are some indications in the Cainozoic record that the major structural features of the region, notably the Wealden dome and the London Basin, became more accentuated during the

Quaternary (Heath, 1984). In other words, some areas underwent uplift and others subsidence whose net effect would depend on contemporaneous shifts of sea level.

Stability (Attewell and Taylor, 1984) is at its most critical in the assessment of potential sites for the disposal of high-level nuclear waste. Some investigators consider the first requirement in the case of continental areas, apart from minimal predictable permeability, is a record suggesting no potential disturbance or change in permeability for the next 1 000 000 years. At sea there is growing interest in subduction zones, where low thermal gradients and low temperatures would ensure little corrosion of containers, and the horst-and-graben structures found in many trench environments would supply repositories for waste subsequently covered by sediments and driven beneath the overriding plate at rates of 2-10 cm/year (Fyfe *et al.*, 1984).

Perhaps the most important attribute of the rock mass under consideration is its permeability, which in turn tends to depend on the extent of cracks and faults. Recent tectonic history, already valuable in the analysis of likely deformation, is here a useful guide to the pattern of fracturing and may reveal that, for example, a granite in orogenic Japan is more promising than an older granite in stable Sweden. In Japan, where the main source of stress is tectonic, cracks in the granite will be steeply dipping conjugate pairs and may be filled by clays; moreover, *in situ* measurements indicate a ratio of horizontal to vertical stresses of $0 \cdot 7$-$1 \cdot 4$ and low vertical movements. In Sweden horizontal sheet cracks may be found to depths of several hundred metres, and uplift can, of course attain 10 mm/year. In fact, this rebound can lead to further cracking. The record of the last 5 million years will also give some indication of climatic changes likely to perturb the tectonic pattern and to remove or deposit cover sediments.

Direct control over future earth movements would seem most plausible where human activity serves to promote or trigger deformation.

There is abundant circumstantial evidence that certain large reservoirs have raised the level of seismic activity nearby. At Lake Mead, on the Colorado River, the number of shocks rises and falls with water level. Filling of the Koyna dam in India began in 1962. There ensued a series of tremors which included one in 1967 with a magnitude of $6 \cdot 4$, and the association between the dam and the seismicity seems well attested (Fig. 110), especially as low reservoir levels appear to account for much reduced seismicity between October 1968 and August 1973 (Gupta and Rastogi, 1974). But there are many more large reservoirs which have apparently failed to produce any such effect (Bolt *et al.*, 1975, p. 30), and a few, such as the Grand Canyon reservoir in Arizona and the Flaming Gorge reservoir in

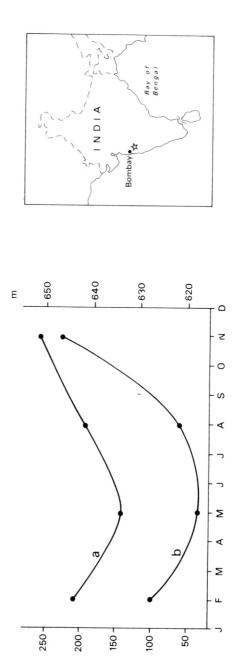

Figure 110. Three-monthly average of water levels (a, scale on right) and numbers of earthquakes (b, scale on left) at Koyna 1964-8. The two trends are closely correlated ($r = 0.93$). Earthquakes occur when high water levels are reached and maintained. After Gupta and Rastogi (1974) with permission. ©1974 by Macmillan Journals Ltd.

Utah, where filling appears to have reduced drastically the incidence of earthquakes within 40 km of the dams (*New Scientist* 1 July 1971, p. 9).

Whether the principal cause is an increase in crustal loading or the leakage of water, the risk is sufficient for site engineers — and prospective litigants — to yearn for good information. Yet there is no example for which the seismic history is sufficiently detailed for firm conclusions to be drawn (Open University, 1981, p. 44). This is understandable: at Koyna, for example, seismographs were not installed until 1963 and the earlier part of the record thus omits minor tremors.

A full narrative is also required to isolate the human contribution where the event at issue is fault movement rather than seismicity. In 1963 the Baldwin Hills reservoir, in Los Angeles, failed when movement on the Reservoir Fault, a subsidiary of the Inglewood Fault, led to seepage through the reservoir lining and in the end to breaching of the dam (Hamilton and Meehan, 1971). A tectonic explanation was rejected. First, there was

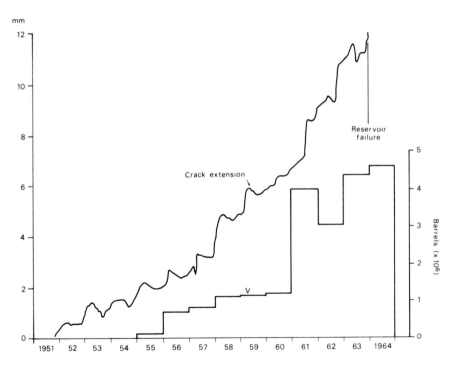

Figure 111. Extension of the Reservoir Fault crack in the inspection chamber of Baldwin Hills Reservoir near Los Angeles in relation to the volume of fluid injected in wells near the fault (v). After Hamilton and Meehan (1971) with the permission. © 1971 The American Association for the Advancement of Science.

no corresponding creep, rupture or strain on the major faults nearby. Second, no regional uplift could be detected in the geodetic data of the last 50 years. Moreover the Inglewood anticline, which is crossed by the Inglewood and Reservoir Faults, shows no evidence of outward movement and if anything the reverse. Much more plausible was the view that fluid injection, practised in the Inglewood oilfield to improve oil recovery, eliminated shear resistance along potential failure planes and led to fault reactivation (Fig. 111). If seismicity and creep cannot be predicted, let alone prevented, controlled release of stress becomes a tempting proposition. But the identification of zones exposed to different levels of risk is clearly preferable, as the dangers can be avoided or guarded against without adding to their unpredictability. And in making predictions the study of recent earth movements graduates from being a descriptive historical field to the status of an experimental science.

REFERENCES

Abdallah, A., Courtillot, V., Kasser, M., Le Dain, A.-Y., Lepine, J. Cl., Robineau, B., Ruegg, J.-C., Tapponnier, P. and Tarantola, A. (1979). Relevance of Afar seismicity and volcanism to the mechanics of accreting plate boundaries. *Nature* **282**, 17-23.

Adams, C. J. (1981). Uplift rates and thermal structure in the Alpine Fault Zone and Alpine Schists, Southern Alps, New Zealand. In: *Thrust and Nappe Tectonics*, pp. 211-222. (McClay, K. R. and Price, N. J., Eds). Blackwell, Oxford.

Adams, F. D. (1938). *The Birth and Development of the Geological Sciences* (republished 1954). Dover, New York.

Adams, J. (1980). Paleoseismicity of the Alpine fault seismic gap, New Zealand. *Geology* **8**, 72-76.

Ager, D. V. (1973). *The Nature of the Stratigraphical Record.* Macmillan, London.

Allen, C. R. (1975). Geological criteria for evaluating seismicity. *Bull. Geol. Soc. Am.* **86**, 1041-1057.

Ambraseys, N. N. (1971). Value of historical records of earthquakes. *Nature* **232**, 357-379.

Ambraseys, N. N. (1978). Studies in historical seismicity and tectonics. In: *The Environmental History of the Near and Middle East since the Last Ice Age*, pp. 185-212. (Brice, W. C., Ed.), Academic Press, London.

Ambraseys, N. N. and Melville, C. P. (1982). *A History of Persian Earthquakes.* Cambridge University Press, Cambridge.

Ambraseys, N. N. and Melville, C. P. (1983). Seismicity of Yemen. *Nature* **303**, 321-323.

Andrews, J. T. (1970). *A Geomorphological Study of Post-glacial Uplift with Particular Reference to Arctic Canada.* Institute of British Geographers, London.

Angelier, J. (1976). La néotectonique cassante et sa place dans un arc insulaire: l'arc égéen méridional. *Rev Géog. Phys. Géol. Dyn.* **18**, 1257-1265.

Angelier, J. (1979), Recent Quaternary tectonics in the Hellenic Arc: examples of geological observations on land. *Tectonophysics* **52**, 269-275.

Argand, E. (1924). La tectonique de l'Asie. *Proc. 13 Int. Geol. Cong.*, 171-372.

Attewell, P. B. and Taylor, R. K. (1984). *Ground Movements and their Effects on Structures*. Surrey University Press, Guildford.

Auden, J. B. (1970). Book review. *Geog. J.* **136**, 288-289.

Babbage, C. (1847). Observations on the Temple of Serapis, at Pozzuoli, near Naples. *Q. J. Geol. Soc. Lond.* **3**, 186-217.

Badgley, P. C. (1965). *Structural and Tectonic Principles*. Harper and Row, New York.

Bailey, E. (1962). *Charles Lyell*. Nelson, London.

Bailey, G. and King, G. C. P. (1985). Tectonics and aggradation in Epirus, Greece. *Proc. Prehist. Soc.* (in press).

Beavan, J., Hauksson, E., McNutt, S. R., Bilham, R. and Jacob, K. H. (1983). Tilt and seismicity changes in the Shumagin Seismic Gap. *Science* **222**, 322-325.

Bellaiche, G. and Blanpied, C. (1979). Aperçu néotectonique, *Géol. Medit.* **6**, 50-59.

Beloussov, V. V. (1962). *Basic Problems in Geotectonics* (Russian original, Moscow 1954, revised). McGraw-Hill, New York.

Beloussov, V. V. (1970). Against the hypothesis of ocean-floor spreading. *Tectonophysics* **9**, 489-511.

Ben Ayed, N., Bobier, Cl., Paskoff, R., Oueslati, A. and Viguier, Cl. (1978). Sur la tectonique récente de la plage du R'mel, à l'Est de Bizerte (Tunisie nord-orientale). *Géol. Medit.* **6**, 423-426.

Bendefy, L. (1968). A method for the elimination of the reference point and of the two different network-adjustments in investigations of recent crustal movements. *Ann. Acad. Sci. Fenn.* **3**, 48-55.

Bender, F. (1974). Geology of Jordan. *Beitr. Reg. Geol. Erde* Suppl. 7.

Bender, P. L. and Silverberg, E. C. (1975). Present tectonic-plate motions from lunar ranging. *Tectonophysics* **29**, 1-7.

Berberian, M. (1976). Contribution to the seismotectonics of Iran (Part II). *Rep. Geol. Surv. Iran* **39**, 1-516.

Bibby, G. (1970), *Looking for Dilmun*. Collins, London.

Bibby, H. M. (1981). Crustal strain from triangulation in Marlborough, New Zealand. *Tectonophysics* **29**, 529-540.

Bilham, R. G. and Beavan, R. J. (1979). Strains and tilts on crustal blocks. *Tectonophysics* **52**, 121-138.

Bilham, R. and Simpson, D. (1984). Indo-Asian convergence and the 1913 survey line connecting the Indian and Russian triangulation surveys. In: *The International Karakoram Project*, (Miller, K. J., Ed.), Vol. I, pp. 160-170. Cambridge University Press, Cambridge.

Bloom, A. L. (1980). Late Quaternary sea level change on South Pacific coasts: a study in tectonic diversity. In: *Earth Rheology, Isostasy and Eustasy* (Mörner, N.-A, Ed.), pp. 505-516. Wiley, Chichester.

Bolt, B. A. (1978). *Earthquakes. A Primer*. Freeman, San Francisco.

Bolt, B. A., Horn, W. L., Macdonald, G. A. and Scott, R. F. (1975). *Geological Hazards*. Springer-Verlag, Berlin.

Bolt, B. A. and Marion, W. C. (1966). Instrumental measurement of slippage on the Hayward Fault. *Bull. Seism. Soc. Am.* **56**, 305-316.

Boore, D. M. (1977). The motion of the ground in earthquakes. *Scient. Am.* **237**, 69-78.

Bourcart, J. (1949). *La Géographie du Fond des Mers*. Payot, Paris.

Bradley, W. C. and Griggs, G. B. (1976). Form, genesis, and deformation of central California wave-cut platforms. *Bull. Geol. Soc. Am.* **87**, 433-449.

Brooks, C. E. P. (1926). *Climate through the Ages*. Benn, London.

Browman, D. L. (1983). Tectonic movement and agrarian collapse in prehispanic Peru. *Nature* **302**, 568-569.

Bucher, W. H. (1933). *The Deformation of the Earth's Crust*. Princeton University Press, Princeton.

Bucknam, R. C. and Anderson, R. E. (1978). Estimation of fault-scarp ages from a scarp-height–slope-angle relationship. *Geology* **7**, 11-14.

Bune, V. I. (1976). A seismic-activity map of the USSR. In: *Seismic Zoning of the USSR* (Medvedev, S. V., Ed.), pp. 87-100. Israel Program for Scientific Translations, Jerusalem.

Burbank, D. W. and Johnson, G. D. (1982). Intermontane-basin development in the past 4 M yr in the north-west Himalaya. *Nature* **298**, 432-436.

Burdon, D. J. (1959). *Handbook of the Geology of Jordan*. Government of the Hashemite Kingdom of Jordan, Amman.

Burford, R. O. (1977). Bimodal distribution of creep event amplitudes on the San Andreas Fault, California. *Nature* **268**, 424-426.

Burnett, A. W. and Schumm, A. W. (1983). Alluvial-river response to neotectonic deformation in Louisiana and Mississippi. *Science* **222**, 49-50.

Camp, L. S. de (1970). *Lost Continents* (1st edn 1954). Dover, New York.

Carey, E. (1979). Recherche des directions principales de contraintes associées au jeu d'une population de failles. *Rev Géog. Phys. Géol. Dyn.* **21**, 57-66.

Carey, S. W. (1976). *The Expanding Earth*. Elsevier, Amsterdam.

Casertano, L., Oliveri del Castillo, A. and Quagliarello, M. T. (1976). Hydrodynamics and geodynamics in the Phlegraean fields area of Italy. *Nature* **264**, 161-164.

Castany, G. (1955). Plissements quaternaires dans la région de Gafsa et le sud Tunisien. *Geol. Rdsch.* **43**, 196-203.

Castle, R. O., Church, J. P., Elliott, M. R. and Morrison, L. (1975). Vertical crustal movements preceding and accompanying the San Fernando earthquake of February 9, 1971: a summary. *Tectonophysics* **29**, 127-140.

Castle, R. O., Church, J. P. and Elliott, M. R. (1976). Aseismic uplift in southern California. *Science* **192**, 251-253.

Cathles, L. M. (1975). *The Viscosity of the Earth's Mantle*. Princeton University Press, Princeton.

Chadwick, W. W., Jr, Swanson, D. A., Iwatsubo, E. Y., Heliker, C. C. and Leighley, T. A. (1983). Deformation monitoring at Mount St Helens in 1981 and 1982. *Science* **221**, 1378-1380.

Chappell, J. (1974). Upper mantle rheology in a tectonic region: evidence from New Guinea. *J. Geophys. Res.* **79**, 456-464.

Charlesworth, J. K. (1957). *The Quaternary Era* (2 vols), Arnold, London.

Chorley, R. J. (1963). Diastrophic background to twentieth-century geomorphological thought. *Bull. Geol. Soc. Am.* **74**, 953-970.

Chorley, R. J., Dunn, A. J. and Beckinsale, R. P. (1964). *The History of the Study of Landforms*, Vol. I. Methuen, London.

Churchill, D. M. (1965). The displacement of deposits formed at sea-level 6500 years ago in Southern Britain. *Quaternaria* **7**, 239-249.

Clark, J. A. (1980). A numerical model of worldwide sea level changes on a viscoelastic earth. In: *Earth Rheology, Isostasy and Eustasy* (Mörner, N.-A., Ed.), pp. 525-534. Wiley, Chichester.

Clark, M. M., Grantz, A. and Rubin, M. (1972). Holocene activity of the Coyote Creek Fault as recorded in sediments of Lake Cahuilla. *US Geol. Surv. Prof. Paper* **787**, 112-130.

Clark, R. H. and Persoage, N. P. (1970). Some implications of crustal movement in engineering planning. *Can. J. Earth Sci.* **7**, 628-633.

Cohen, S. C. (1984). Postseismic deformation due to subcrustal viscoelastic relaxation following dip-slip earthquakes. *J. Geophys. Res.* **89**, 4538-4544.

Coles, J. M. and Higgs, E. S. (1969). *The Archaeology of Early Man*. Faber and Faber, London.

Coque, R. (1962). *La Tunisie Présaharienne*. Colin, Paris.

Cressman, G. P. (1961). Weather prediction—dynamic in daily use. In: *The Encyclopedia of Atmospheric Sciences and Astrogeology*, (Fairbridge, R. W., Ed.), pp. 1123-1133. Reinhold, New York.

Crittenden, M. D., Jr (1963). Effective viscosity of the earth derived from isostatic loading of Pleistocene Lake Bonneville. *J. Geophys. Res.* **68**, 5517-5530.

Crittenden, M. D., Jr (1967a). New data on the isostatic deformation of Lake Bonneville. *US Geol. Surv. Prof. Paper* **454E**, 1-31.

Crittenden, M. D., Jr (1967b). Viscosity and finite strength of the mantle as determined from water and ice loads. *Geophys. J. Roy. Astron. Soc.* **14**, 261-279.

Daly, R. A. (1926). *Our Mobile Earth*. Charles Scribner's Sons, New York.

Daly, R. A. (1934). *The Changing World of the Ice Age*. Yale University Press, New Haven.

Davis, W. M. (1899). The geographical cycle. *Geog. J.* **14**, 481-504.

Davis, W. M. (1909). *Geographical Essays*. Ginn, Boston.

Davison, C. (1937). Founders of seismology, IV. *Geol. Mag.* **74**, 529-534.

Decker, R. W. and Kinoshita, W. T. (1972). Geodetic measurements. In: *The Surveillance and Prediction of Volcanic Activity*, pp. 47-74. UNESCO, Paris.

Decker, R. W., Einarsson, P. and Mohr, P. A. (1971). Rifting in Iceland: new geodetic data. *Science* **173**, 530-533.

Deng, O., Fengmin, S., Shilong, Z., Mengluan, L., Tielin, W., Weiqi, Z., Burchfel, B. S., Molnar, P. and Peizhen, Z. (1984). Active faulting and tectonics of the Ningxia-Hui Autonomous Region, China. *J. Geophys. Res.* **89**, 4427-4445.

Denham, D., Alexander, L. G. and Worotnicki, G. (1980). The stress field near the sites of the Meckering (1968) and Calingiri (1970) earthquakes, Western Australia. *Tectonophysics* **67**, 283-317.

Deschamps, A. and King, G. C. P. (1983). The Campania-Lucania (southern Italy) earthquake of 23 November 1980. *Earth Planet. Sci. Lett.* **62**, 296-304.

Donner, J. (1980). The determination and dating of synchronous Late Quaternary shorelines in Fennoscandia. In: *Earth Rheology, Isostasy and Eustasy* (Mörner, N.-A., Ed.), pp. 285-293. Wiley, Chichester.

Doornkamp, J. 'C., Brunsden, D. and Jones, D. K. C. (Eds) (1980). *Geology, Geomorphology and Pedology of Bahrain.* Geobooks, Norwich.

Downey, W. S. and Tarling, D. H. (1984). Archaeomagnetic dating of Santorini volcanic eruptions and fired destruction levels of late Minoan civilization. *Nature* **309**, 519-523.

Dreghorn, W. (1981). Recent uplift in northern Cyprus. *Geol. en Mijnb.* **60**, 281-284.

Du Toit, A. L. (1937). *Our Wandering Continents.* Oliver and Boyd, Edinburgh.

Dvorak, J., Okamura, A. and Dieterich, J. H. (1983). Analysis of surface deformation data, Kilauea Volcano, Hawaii, October 1966 to September 1970. *J. Geophys. Res.* **88**, 9295-9304.

Edge, R. J., Baker, T. F. and Jeffries, G. (1981). Borehole tilt measurements: aperiodic crustal tilt in an aseismic area. *Tectonophysics* **71**, 97-109.

Eronen, M. (1983). Late Weichselian and Holocene shore displacement in Finland. In: *Shorelines and Isostasy* (Smith, D. E. and Dawson, A. G., Eds), pp. 183-207. Academic Press, London.

Falcon, N. L. (1974). Southern Iran: Zagros Mountains. In: *Mesozoic-Cenozoic Orogenic Belts* (Spencer, A. M., Ed.), pp. 199-211. Scottish Academic Press, Edinburgh.

Farhoudi, G. and Karig, D. E. (1977). Makran of Iran and Pakistan as an active arc system. *Geology* **5**, 664-668.

Faure, H., Fontes, J. C., Hébrard, L., Monteillet, J. and Pirazzoli, P. A. (1980). Geoidal change and shore-level tilt along Holocene estuaries: Sénégal River area, West Africa. *Science* **210**, 421-423.

Finetti, I. and Morelli, C. (1972). Wide scale digital seismic exploration of the Mediterranean Sea. *Boll. Geofis. Teorica Appl.* **14**, 291-342.

Flemming, N. C. (1969). Archaeological evidence for eustatic change of sea level and earth movements in the western Mediterranean during the last 2,000 years. *Geol. Soc. Am. Spec. Pap.* **109**, 1-125.

Flemming, N. C. (1971). Eustatic and tectonic factors in the relative vertical displacement of the Aegean coast. In: *The Mediterranean Sea* (Stanley, D. J., Ed.), Dowden, Hutchinson and Ross, Stroudsburg.

Flemming, N. C. (1978). Holocene eustatic changes and coastal tectonics in the northeast Mediterranean: implications for models of crustal consumption. *Phil. Trans. Roy. Soc. Lond.* **289** A, 405-458.

Flemming, N. C., Raban, A. and Goetschel, C. (1978). Tectonic and eustatic changes in the Mediterranean coast of Israel in the last 9000 years. *Prog. Underwater Sci.* **3**, 33-93.

Flint, R. F. (1970). *Glacial and Quaternary Geology.* Wiley, New York.

Forbes, R. J. (1963). *Studies in Ancient Technology* Vol. VIII. Brill, Leiden.

Francheteau, J. (1983). The oceanic crust. *Scient. Am.* **249**, 68-84.

Freund, R. (1965). A model of the structural development of Israel and adjacent areas since Upper Cretaceous times. *Geol. Mag.* **102**, 189-205.

Fyfe, W. F., Babuska, V., Price, N. J., Schmid, E., Tsang, C. F., Uyeda, S. and Velde, B. (1984). The geology of nuclear waste disposal. *Nature* **310**, 537-540.

Gansser, A. (1964). *Geology of the Himalayas.* Interscience, London.

Garfunkel, Z., Zak, I. and Freund, R. (1981). Active faulting in the Dead Sea Rift. *Tectonophysics* **80**, 1-26.

Gerard, V. B. (1975). Possible large creep event apparently preceded by a dilatant precursor. *Nature* **253**, 520.

Gerasimov, I. P. (Ed.) (1967). *Recent Crustal Movements* (Russian original, Moscow 1963). Israel Program for Scientific Translations, Jerusalem.

Gerke, K. (1974). Crustal movements in the Myvatu — and in the Thingvallavatn — area, both horizontal and vertical. In: *Geodynamics of Iceland and the North Atlantic Area* (Kristjansson, L., Ed.), pp. 263-275. Reidel, Dordrecht.

Gibbs, A. D. (1983). Balanced cross-section construction from seismic sections in areas of extensional tectonics. *J. Struct. Geol.* **5**, 153-160.

Gilbert, G. K. (1890). Lake Bonneville. *Monogr. US Geol. Surv.* 1-438.

Girdler, R. W. and Styles, P. (1978). Seafloor spreading in the Western Gulf of Aden. *Nature* **271**, 615-617.

Goguel, J. (1962). *Tectonics.* W. H. Freeman, San Francisco.

Gordon, F. R. (1971). Faulting during the earthquake at Meckering, Western Australia: 14 October 1968. *Bull. Roy. Soc. NZ* **9**, 85-93.

Gorshkov, G. and Yakushova, A. (1967). *Physical Geology.* Mir, Moscow.

Gough, D. I. (1984). Mantle upflow under North America and plate dynamics. *Nature* **311**, 428-433.

Gough, D. I., Fordjor, C. K. and Bell, J. S. (1983). A stress province boundary and tractions on the North American plate. *Nature* **305**, 619-621.

Gould, S. J. (1965). Is uniformitarianism necessary? *Am. J. Sci.* **263**, 223-228.

Gould, S. J. (1980). *The Panda's Thumb.* Norton, New York.

Grant, D. R. (1970). Recent coastal submergence of the Maritime Provinces, Canada. *Can J. Earth Sci.* **7**, 676-689.

Grant, D. R. (1980). Quaternary sea-level change in Atlantic Canada as an indication of crustal delevelling. In: *Earth Rheology, Isostasy and Eustasy* (Mörner, N.-A., Ed.), pp. 201-214. Wiley, Chichester.

Grant-Taylor, T. L. *et al.* (1974). Microzoning for earthquake effects in Wellington N. Z. *NZ DSIR Bull.* **213**, 1-62.

Greenwood, G. (1857). *Rain and Rivers.* London.

Greiner, G. and Illies, J. H. (1977). Central Europe: active or residual tectonic stresses. In: *Stress in the Earth* (Wyss, M., Ed.), pp. 11-26. Birkhäuser, Basel.

Grob, H., Kovari, K. and Amstad, C. (1975). Sources of error in the determination of in-situ stresses by measurements. *Tectonophysics* **29**, 29-39.

Guest, J. (1982). Sir William Hamilton — pioneer volcanologist. *Earthquake Inform. Bull* **14**, 48-55.

Gupta, H. K. and Rastogi, B. K. (1974). Will another damaging earthquake occur in Koyna? *Nature* **248**, 215-216.

Gutenberg, B. and Richter, C. G. (1949). *Seismicity of the Earth* (2nd edn 1954). Princeton University Press, Princeton.

Haimson, B. C. and Voight, B. (1977). Crustal stress in Iceland. In: Stress in the Earth (Wyss, M., Ed.), pp. 153-198. Birkhäuser, Basel.

Hallam, A. (1973). *A Revolution in the Earth Sciences.* Oxford University Press, Oxford.

Hallam, A. (1983). *Great Geological Controversies.* Oxford University Press, Oxford.

Hamblin, W. K. (1984). Direction of absolute movement along the boundary faults of the Basin and Range — Colorado Plateau margin. *Geology* **12**, 116-119.

Hamilton, D. H. and Meehan, R. L. (1971). Ground rupture in the Baldwin Hills. *Science* **172**, 333-344.

Hancock, P. L. and Barka, A. A. (1980). Plio-Pleistocene reversal of displacement on the North Anatolian Fault Zone. *Nature* **286**, 591-594.

Heath, M. (1984). A comment upon the paper by John and Fisher, P. G. A., 95/3. *Proc. Geol. Assoc. Circ.* **846**, 10.

Henbest, N. (1984). Continental drift: the final proof. *New Scientist* **105**, 6.

Hobbs, B. E., Means, W. D. and Williams, P. F. (1976). *An Outline of Structural Geology.* Wiley, New York.

Hollingworth, S. E. (1964). Dating the uplift of the Andes of northern Chile. *Nature* **201**, 17-20.

Holmes, A. (1965). *Principles of Physical Geology* (2nd edn). Nelson, London.

Hooke, R. (1705). *The Posthumous Works of Dr. Robert Hooke.* Smith and Walford, London.

Horai, K. (1982). A satellite altimetric geoid in the Philippine Sea. *Nature* **299**, 117-121.

Horowitz. A. (1979). *The Quaternary of Israel.* Academic Press, New York.

Howell, B. F., Jr (1973). Average regional siesmic hazard index (ARSHI) in the United States. In: *Geology, Seismicity, and Environmental Impact* (Moran *et al.*, Eds), pp. 277-285. University Publishers, Los Angeles.

Hsü, K. J. (1983). Uniformitarianism under scrutiny. *Geology* **11**, 683-684.

Huntoon, P. W. (1977). Holocene faulting in the Western Grand Canyon, Arizona. *Bull. Geol. Soc. Am.* **88**, 1619-1622.

Isaac, G. Ll. (1972). Chronology and the tempo of cultural change during the Pleistocene. In: *Calibration of Hominoid Evolution* (Bishop, W. W. and Miller, J. A., (Eds), pp. 381-430. Scottish Academic Press, Edinburgh.

Isachsen, Y. W. (1975). Possible evidence for contemporary doming of the Adirondack Mountains, New York, and suggested implications for regional tectonics and seismicity. *Tectonophysics* **29**, 169-181.

Isacks, B., Oliver, J. and Sykes, L. R. (1968). Seismology and the new global tectonics. *J. Geophys. Res.* **73**, 5855-5899.

Jackson, J. and McKenzie, D. (1983). The geometrical evolution of normal fault systems. *J. Struct. Geol.* **5**, 471-482.

Jackson, J. A. and McKenzie, D. P. (1984). Active tectonics of the Alpine-Himalayan Belt between western Turkey and Pakistan. *Geophys. J. Roy. Astron. Soc.* **77**, 185-264.

Jackson, J. A., Gagnepain, J., Houseman, G., King, G. C. P., Papadimitriou, P., Soufleris, C. and Virieux, J. (1982a). Seismicity, normal faulting, and the geomorphological development of the Gulf of Corinth (Greece): the Corinth earthquakes of February and March 1981. *Earth Planet. Sci. Lett.* **57**, 377-397.

Jackson, J. A., King, G. and Vita-Finzi, C. (1982b). The neotectonics of the Aegean: an alternative view. *Earth Planet. Sci. Lett.* **61**, 303-318.

Jeffreys, H. (1942). On the mechanisms of faulting. *Geol. Mag.* **89**, 291-295.

Jeffreys, H. (1970). *The Earth* (5th edn). Cambridge University Press, Cambridge.

Jeffreys, H. (1975). The Fenno-Scandian uplift. *J. Geol. Soc. Lond.* **131**, 323-325.

Johnston, M. J. S., McHugh, S. and Burford, R. O. (1976). On simultaneous tilt and creep observations on the San Andreas Fault. *Nature* **260**, 691-693.

Kachadoorian, R. (1968). Effects of the earthquake of March 27, 1964, on the Alaska highway system. *US Geol. Surv. Prof. Paper* **545-C**.

Kanamori, H. (1973). Mode of strain release associated with major earthquakes in Japan. *Annu. Rev. Earth Planet. Sci.* **1**, 213-239.

Kanamori, H. (1978). Quantification of earthquakes. *Nature* **271**, 411-414.

Kanamori, H. (1983). Global seismicity. In: *Earthquakes: Observation, Theory and Interpretation* (Kanamori, H. and Boschi, E., Eds), pp. 596-608. North-Holland, Amsterdam.

Karcz, I. and Kafri, U. (1973). Recent vertical crustal movements between the Dead Sea Rift and the Mediterranean. *Nature* **242**, 42-43.

Karcz, I. and Kafri, U. (1975). Recent crustal movements along Mediterranean coastal plain of Israel. *Nature* **257**, 296-297.

Kasahara, K. (1971). The role of geodesy in crustal movement studies. *Bull. Roy. Soc. NZ* **9**, 1-5.

Kasahara, K. (1981). *Earthquake Mechanics.* Cambridge University Press, Cambridge.

Kassler, P. (1973). The structural and geomorphic evolution of the Persian Gulf. In: *The Persian Gulf* (Purser, B., Ed.), pp. 11-32. Springer-Verlag, Berlin.

Kaufman, A. and Broecker, W. (1965). Comparison of Th^{230} and C^{14} ages for carbonate materials from Lakes Lahontan and Bonneville. *J. Geophys. Res.* **70**, 4039-4054.

Kaula, W. M. (1980). Problems in understanding vertical movements and earth rheology. In: *Earth Rheology, Isostasy and Eustasy* (Mörner, N.-A., Ed.), pp. 577-588. Wiley, Chichester.

Kaula, W. M. (1983). The changing shape of the Earth. *Nature* **303**, 756.

Kelsey, J. (1972). Geodetic aspects concerning possible subsidence in southeastern England. *Phil. Trans. Roy. Soc. Lond.* **272**, 141-150.

Kerr, R. A. (1984). An encouraging long-term quake "forecast". *Science* **225**, 300-301.

Kilston, S. and Knopoff, L. (1983). Lunar-solar periodicities of large earthquakes in southern California. *Nature* **304**, 21-25.

King, G. A. M. (1983). Tiltmeter recordings re-analysed for earthquake precursors. *Nature* **302**, 272.

King, G. and Stein, R. (1983). Surface folding, river terrace deformation rate and earthquake repeat time in a reverse faulting environment: the Coalinga, California, earthquake of May 1983. In: *The 1983 Coalinga, California, Earthquake.* Calif. Div. Mines & Geol. Spec. Pub. 66.

King, G., Soufleris, C. and Berberian, M. (1981). The source parameters, surface deformation and tectonic setting of three recent earthquakes: Thessaloniki (Greece), Tabas-e-Golshan (Iran) and Carlisle (UK). *Disasters* **5**, 36-46.

King, G. C. P. and Vita-Finzi, C. (1981). Active folding in the Algerian earthquake of 10 October 1980. *Nature* **292**, 22-26.

King, G. C. P. and Yielding, G. (1984). The evolution of a thrust fault system: processes of rupture initiation, propagation and termination in the 1980 El Asnam, Algeria earthquake. *Geophys. J. Roy. Astron. Soc.* **76**, 915-933.

King, G. C. P., Ouyang, Z. X., Papadimitriou, P., Deschamps, A., Gagnepain, J., Houseman, G., Jackson, J. A., Soufleris, C. and Virieux, J. (1985). The evolution of the Gulf of Corinth (Greece): an aftershock study of the 1981 earthquakes. *Geophys. J. Roy. Astron. Soc.* **80**, 677-693.

King, L. C. (1952). *The Morphology of the Earth.* Oliver and Boyd, Edinburgh.

King, L. C. (1983). *Wandering Continents and Spreading Sea Floors on an Expanding Earth.* Wiley, New York.

King-Hele, D. (1976). The shape of the Earth. *Science* **192**, 1293-1300.

Kisslinger, C. (1978). Seismicity and global tectonics as a framework for microzonation. *Proc. 2nd Int. Conf. Microzonation, San Francisco,* **I**, 3-25.

Kiviniemi, A. (1981). Some results concerning crustal movements in Finland. *Tectonophysics* **24**, 65-71.

Lamar, D. L., Merifield, P. M. and Proctor, R. J. (1973). Earthquake recurrence intervals on major faults in southern California. In: *Geology, Seismicity and Environmental Impact* (Moran, D. E. *et al.*, Eds), pp. 265-276. University Publishers, Los Angeles.

Lambrick, H. T. (1967). The Indus floodplain and the 'Indus' civilization. *Geog. J.* **133**, 493-494.

Lees, G. M. (1955). Recent earth movements in the Middle East. *Geol. Rdsch.* **43**, 221-6.

Lees, G. M. and Falcon, N. L. (1952). The geographical history of the Mesopotamian plains. *Geog. J.* **118**, 24-39.

Lewis, K. B. (1971). Growth rate of folds using tilted wave-planed surfaces: coast and continental shelf, Hawke's Bay, New Zealand. *Bull. Roy. Soc. NZ* **9**, 225-231.

Lipman, P. W. and Mullineaux, D. R. (Eds) (1981). The 1980 eruptions of Mount St. Helens, Washington. *US Geol. Surv. Prof. Paper* **1250**.

Lisitzin, E. (1974). *Sea-level Changes.* Elsevier, Amsterdam.

Lliboutry, L. A. (1971). Rheological properties of the asthenosphere from Fennoscandian data. *J. Geophys. Res.* **76**, 1433-1446.

Lomnitz, C. (1974). *Global Tectonics and Earthquake Risk.* Elsevier, Amsterdam.

Lomnitz, C. and Lomnitz, L. (1978). Tangshan 1976: a case history in earthquake prediction. *Nature* **271**, 109–111.

Lyell, C. (1837). *Principles of Geology* (5th edn). Murray, London.

Lyttleton, R. A. (1982). *The Earth and its Mountains.* Wiley, New York.

Lyustikh, E. N. (1960). *Isostasy and Isostatic Hypotheses* (first published in Russian in *Trudy Geofiz. Inst.* **38**, 1959). Am. Geophys. Union, Washington, DC.

McClure, H. A. and Vita-Finzi, C. (1982). Holocene shorelines and tectonic movements in eastern Saudi Arabia. *Tectonophysics* **85**, T37–43.

McEwin, A., Underwood, R. and Denham, D. (1976). Earthquake risk in Australia. *BMR J. Aus. Geol. Geophys.* **1**, 15–21.

Macfadyen, W. A. and Vita-Finzi, C. (1978). Mesopotamia: the Tigris-Euphrates delta and its Holocene Hammar fauna. *Geol. Mag.* **115**, 287–300.

McKenzie, D. P. (1969). The relation between fault plane solutions for earthquakes and the directions of the principal stresses. *Bull. Seism. Soc. Am.* **59**, 591–601.

McKenzie, D. P. (1972). Active tectonics of the Mediterranean region. *Geophys. J. Roy. Astron. Soc.* **30**, 109–185.

McKenzie, D. (1978). Active tectonics of the Alpine–Himalayan belt: the Aegean Sea and surrounding regions. *Geophys. J. Roy. Astron. Soc.* **55**, 217–254.

McKusick, M. and Shinn, E. A. (1980). Bahamian Atlantis reconsidered. *Nature* **287**, 11–12.

Mann, C. D. and Vita-Finzi, C. (1982). Curve interpolation and folded strata. *Tectonophysics* **88**, 7–15.

Manzoni, M. (1968). *Dizionario di Geologia.* Zanichelli, Bologna.

Marshall, P. D. (1983). Predicting disasters (book review). *Nature* **305**, 251.

Mather, K. F. and Mason, S. L. (Eds) (1939). *A Source Book in Geology.* McGraw-Hill, New York.

Matsuda, T., Ota, Y., Ando, M. and Yonekura, N. (1978). Fault mechanism and recurrence time of major earthquakes in southern Kanto district, Japan, as deduced from coastal terrace data. *Bull. Geol. Soc. Am.* **89**, 1610–1618.

Mattskova, V. A. (1967). A revised velocity map of recent vertical crustal movements in the western half of the European USSR, and some remarks on the period of these movements. In: *Recent Crustal Movements* (Gerasimov, I. P., Ed.), pp. 76–89. Israel Program for Scientific Translations, Jerusalem.

Matuzawa, T. (1964). *Study of Earthquakes.* Uno, Tokyo.

Meade, B. K. (1971). Horizontal movement along the San Andreas Fault System. *Bull. Roy. Soc. NZ* **9**, 175–179.

Medvedev, S. V. (1965). *Engineering Seismology* (Russian original 1962). Israel Program for Scientific Translations, Jerusalem.

Medvedev, S. V. (Ed.) (1976). *Seismic Zoning of the USSR* (Russian original, Moscow 1968). Israel Program for Scientific Translations, Jerusalem.

Melchior, P. (1978). *The Tides of the Planet Earth.* Pergamon, Oxford.

Mercier, J.-L. (1976). La néotectonique, ses méthodes et ses buts. Un example: l'arc egéen (Méditerranée orientale). *Rev Géog. Phys. Géol. Dyn.* **18**, 323-346.

Mercier, J. L., Delibassis, N., Gauthier, A., Jarrige, J.-J., Lemeille, F., Philip, H., Sébrier, M. and Sorel, D. (1979). La néotectonique de l'Arc Egéen. *Rev. Géog. Phys. Géol. Dyn.* **21**, 67-92.

Mescheryakov, Yu. A. (1967). Secular movements of the Earth's crust: some results and problems. In: *Recent Crustal Movements* (Gerasimov, I. P., Ed.), pp. 1-21. Israel Program for Scientific Translations, Jerusalem.

Minster, J. B. and Jordan, T. H. (1978). Present-day plate motions. *J. Geophys. Res.* **83**, 5331-5354.

Miyashiro, A., Aki, K. and Şengör, A. M. C. (1982). *Orogeny.* Wiley, Chichester.

Mjachkin, V. I., Brace, W. F., Sobolev, G. A. and Dieterich, J. H. (1975). Two models for earthquake forerunners. In: *Earthquake Prediction and Rock Mechanics* (Wyss, M., Ed.), pp. 169-181. Birkhäuser, Basel.

Molnar, P. and Chen, W.-P. (1982). Seismicity and mountain building. In: *Mountain Building Processes* (Hsü, K. J., Ed.), pp. 41-57. Academic Press, London.

Molnar, P. and Tapponnier, P. (1975). Cenozoic tectonics of Asia: effects of a continental collision. *Science* **189**, 419-426.

Molnar, P. and Tapponnier, P. (1978). Active tectonics of Tibet. *J. Geophys. Res.* **83**, 5361-5375.

Moran, D. E. *et al.* (Eds) (1973). *Geology, Seismicity, and Environmental Impact.* University Publishers, Los Angeles.

Morgan, W. J. (1968). Rises, trenches, great faults, and crustal blocks. *J. Geophys. Res.* **73**, 1959-1982.

Mörner, N.-A. (Ed.) (1980). *Earth Rheology, Isostasy and Eustasy.* Wiley, Chichester.

Mörner, N.-A. (1980). The Fennoscandian uplift: geological data and their geodynamical implication. In: *Earth Rheology, Isostasy and Eustasy* (Mörner, N.-A., Ed.), pp. 251-283. Wiley, Chichester.

Morrison, R. B. (1965*a*). New evidence on Lake Bonneville stratigraphy and history from Southern Promontory Point, Utah. *US Geol. Surv. Prof. Paper* **525**C, 110-119.

Morrison, R. B. (1965*b*). Quaternary geology of the Great Basin. In: *The Quaternary of the United States* (Wright, H. E., Jr and Frey, D. G., Eds), pp. 265-285. Princeton University Press, Princeton.

Murray, J. B. and Guest, J. E. (1982). Vertical ground deformation on Mount Etna, 1975-1980. *Bull. Geol. Soc. Am.* **93**, 1160-1175.

Murray, J. B., Guest, J. E. and Butterworth, P. S. (1977). Large ground deformation on Mount Etna volcano. *Nature* **266**, 338-340.

Needham, J. (1959). *Science and Civilisation in China,* Vol. 3, Cambridge University Press, Cambridge.

Neev, D. and Friedman, G. M. (1978). Late Holocene tectonic activity along the margins of the Sinai subplate. *Science* **202**, 427-430.

Neev, D., Bakler, N., Moshkovitz, S., Kaufman, A., Magaritz, M. and Gofna, R. (1973). Recent faulting along the Mediterranean coast of Israel. *Nature* **245**, 254-256.

Newman, W. S., Cinquemani, L. J., Pardi, R. R. and Marcus, L. F. (1980). Holocene delevelling of the United States' east coast. In: *Earth Rheology, Isostasy and Eustasy* (Mörner, N.-A., Ed.), pp. 449-463. Wiley, Chichester.

Newmark, R. L., Zoback, M. D. and Anderson, R. N. (1984). Orientation of *in situ* stresses in the oceanic crust. *Nature* **311**, 424-428.

Nikonov, A. A. (1980). Manifestations of glacio-isostatic processes in northern countries during the Holocene and at present. In: *Earth Rheology, Isostasy and Eustasy* (Mörner, N.-A., Ed.), pp. 341-354. Wiley, Chichester.

North, R. G. (1974). Seismic slip rates in the Mediterranean and Near East. *Nature* **252**, 560-563.

Oldham, R. D. (1899). Report on the Great Earthquake of 12th June 1897. *Mem. Geol. Surv. India* **29**, 1-379.

Oldham, R. D. (1926). The Cutch (Kachh) Earthquake of 16th June 1819 with a revision of the Great Earthquake of 12th June 1897. *Mem. Geol. Surv. India* **46**, 71-147.

Ollier, C. D. (1981). *Tectonics and Landforms.* Longman, London. Open University (1981). *Earth Structure*, S237, Block 2. Open University Press, Milton Keynes.

Page, W. D., Alt, J. N., Cluff, L. S. and Plafker, G. (1979). Evidence for the recurrence of large-magnitude earthquakes along the Makran Coast of Iran and Pakistan. *Tectonophysics* **52**, 533-547.

Page, W. D., Anttonen, G. and Savage, W. U. (1978). The Makran coast of Iran, a possible seismic gap. *Proc. Conf. VI Methodology for Identifying Seismic Gaps and Soon-to-break Gaps*, US Geol. Surv. Open-File Rep. 78-943, 611-633.

Papastamatiou, D. (1980). Incorporation of crustal deformation to seismic hazard analysis. *Bull. Seism. Soc. Am.* **70**, 1321-1335.

Paskoff, R. and Sanlaville, P. (1980). Le Tyrrhénienn de la Tunisie. *C.R. Acad. Sci. Paris* **290**, 393-396.

Peltier, R. (1981). Ice age geodynamics. *Annu. Rev. Earth Planet. Sci.* **9**, 199-225.

Peltier, W. R. (1983). Constraint on deep mantle viscosity from Lageos acceleration data. *Nature* **304**, 434-436.

Peltier, W. R. and Andrews, J. T. (1983). Glacial geology and glacial isostasy of the Hudson Bay region. In: *Shorelines and Isostasy* (Smith, D. E. and Dawson, A. G., Eds), pp. 285-319. Academic Press, London.

Penck, W. (1953). *Morphological Analysis of Landforms* (German original, 1924). Macmillan, London.

Picard, L. and Baida, U. (1966). *Geological Report on the Lower Pleistocene Deposits of the 'Ubeidiya Excavations.* Israel Acad. Sci. Hum., Jerusalem.

Pillans, B. (1983). Upper Quaternary marine terrace chronology and deformation, South Taranaki, New Zealand. *Geology* **11**, 292-297.

Pirazzoli, P. (1973). Inondations et niveaux marins à Venise. *Mém Lab. Géomorph. École Pratique Hautes Ét.* **22**, Dinard.

Pirazzoli, P. A., Thommeret, J., Thommeret, Y., Labord, J. and Montaggioni, L. F. (1982). Crustal block movements from Holocene shorelines: Crete and Antikythira (Greece). *Tectonophysics* **86**, 27-43.

Plafker, G. (1965). Tectonic deformation associated with the 1964 Alaska earthquake. *Science* **148**, 1675-1687.

Plafker, G. (1967). Surface faults on Montague Island associated with the 1964 Alaska earthquake. *US Geol. Surv. Prof. Paper* **543 G**, 1-42.

Plafker, G. (1969). Tectonics of the March 27, 1964 Alaska Earthquake. *US Geol. Surv. Prof. Paper* **543-I**, 1-74.

Poirier, J. P., Romanowicz, B. A. and Taher, M. A. (1980). Large historical earthquakes and seismic risk in Northwest Syria. *Nature* **285**, 217-220.

Powell, J. W. (1875). *Exploration of the Colorado River of the West.* Washington.

Price, N. J. (1975). Rates of deformation. *J. Geol. Soc. Lond.* **131**, 553-575.

Prilepin, M. T. (1981). The present state and prospects of development of high-precision geodetic methods for studying recent crustal movements. *Tectonophysics* **71**, 13-25.

Quennell, A. M. (1958). The structural and geomorphic evolution of the Dead Sea Rift. *Q. J. Geol. Soc.* **114**, 1-24.

Radbruch, D. (1971). Progress in planning and regulation of construction in active fault zones, San Francisco Bay area, California. *Bull. Roy. Soc. NZ* **9**, 189-193.

Raikes, R. L. (1964). The end of the ancient cities of the Indus. *Am. Anthrop.* **66**, 284-299.

Raikes, R. L. (1965). The Mohenjo-daro Floods. *Antiquity* **39**, 196-203.

Raikes, R. L. (1967). *Water, Weather and Prehistory.* Baker, London.

Ramsay, J. G. (1967). *Folding and Fracturing of Rocks.* McGraw-Hill, New York.

Ranalli, G. (1975). Geotectonic relevance of rock stress determinations. *Tectonophysics* **29**, 49-56.

Reches, Z. and Hoexter, D. F. (1981). Holocene seismic and tectonic activity in the Dead Sea area. *Tectonophysics* **80**, 235-254.

Richards, G. W. (1982). *Mediterranean Intertidal Molluscs as Sea-level Indicators.* Ph.D. thesis, London University.

Richter, C. F. (1958). *Elementary Seismology.* W. H. Freeman, San Francisco.

Rikitake, T. (1976). Recurrence of great earthquakes at subduction zones. *Tectonophysics* **35**, 335-362.

Robbins, A. R. (1980). Introduction. *Phil. Trans. Roy. Soc. Lond.* **294** A, 211-215.

Rossiter, J. R. (1972). Sea-level observations and their secular variation. *Phil. Trans. Roy. Soc. Lond.* **272** A, 131-139.

Sacks, I. S., Suyehiro, S., Linde, A. T. and Snoke, J. A. (1978). Slow earthquakes and stress redistribution. *Nature* **275**, 599-602.

Sauramo, M. (1939). The mode of land upheaval in Fennoscandia during Late-Quaternary time. *C.R. Soc. Géol. Finlande* **13**, 1-26.

Schaer, J. P., Reimer, G. M. and Wagner, G. A. (1981). Actual and ancient uplift rate in the Gotthard region, Swiss Alps: a comparison between precise levelling and fission-track apatite age. *Tectonophysics* **24**, 293-300.

Scheidegger, A. E. (1963). *Principles of Geodynamics* (2nd edn). Springer-Verlag, Berlin.

Scheidegger, A. E. (1970). *Theoretical Geomorphology* (2nd edn). Allen and Unwin, London.

Scheidegger, A. E. (1975). *Physical Aspects of Natural Catastrophes.* Elsevier, Amsterdam.

Scholz, C. H., Sykes, L. R. and Aggarwal, Y. P. (1973). Earthquake prediction: a physical basis. *Science* **181**, 803-810.

Schulz, S., Burford, R. O. and Mavko, B. (1983). Influence of seismicity and rainfall on episodic creep on the San Andreas Fault System in Central California. *J. Geophys Res.* **88**, 7475-7484.

Schumm, S. A. (1963). The disparity between present rates of denudation and orogeny. *U.S. Geol. Surv. Prof. Paper* **454** H.

Schwartz, D. P. and Coppersmith, K. J. (1984). Fault behavior and characteristic earthquakes: examples from the Wasatch and San Andreas fault zones. *J. Geophys. Res.* **89**, 5681-5698.

Shakal, A. G. and Toksöz, M. N. (1977). Earthquake hazard in New Zealand. *Science* **195**, 171-173.

Shapiro, I. I. (1983). Use of space techniques for geodesy. In: *Earthquakes: Observation, Theory and Interpretation,* (Kanamori, H. and Boschi, E., Eds), pp. 530-568. North-Holland, Amsterdam.

Shearman, D. J. (1976). The geological evolution of Southern Iran. *Geog. J.* **142**, 393-410.

Shennan, I. (1983). Flandrian and Late Devensian sea-level changes and crustal movements in England and Wales. In: *Shorelines and Isostasy* (Smith, D. E. and Dawson, A. G., Eds), pp. 255-283. Academic Press, London.

Sibson, R. H., White, S. H. and Atkinson, B. K. (1981). Structure and distribution of fault rocks in the Alpine Fault Zone, New Zealand. In: *Thrust and Nappe Tectonics* (McClay, K. R. and Price, N. J., Eds), pp. 197-210. Blackwell, Oxford.

Sieh, K. (1978*a*). Prehistoric large earthquakes produced by slip on the San Andreas Fault at Pallett Creek, California. *J. Geophys Res.* **83**, 3907-3939.

Sieh, K. E. (1978*b*). Slip along the San Andreas Fault associated with the great 1857 earthquake. *Bull. Seism. Soc. Am.* **68**, 1421-1448.

Sieh, K. E. (1981). A review of geological evidence for recurrence times of large earthquakes. *Earthquake Prediction (Maurice Ewing Series)* **4**, 181-207.

Sieh, K. E. (1984). Lateral offsets and revised dates of large prehistoric earthquakes at Pallett Creek, Southern California. *J. Geophys. Res.* **89**, 7641-7670.

Sieh, K. E. and Jahns, R. H. (1984). Holocene activity of the San Andreas fault at Wallace Creek, California. *Bull. Geol. Soc. Am.* **95**, 883-896.

Siirinäinen, A. (1972). A gradient/time curve for dating Stone Age shorelines in Finland. *Suomen Museo*, 5-18.

Simpson, D. W. (1980). Earthquake prediction. *Nature* **286**, 445-456.

Sissons, J. B. and Cornish, R. (1982). Rapid localized glacio-isostatic uplift at Glen Roy, Scotland. *Nature* **297**, 213-214.

Slemmons, D. B. (1977). Faults and earthquake magnitude. *US Corps of Engineers, Waterways Exp. Station Misc. Pap. S-73-1.*

Smalley, I. J. (1967). The subsidence of the North Sea Basin and the geomorphology of Britain. *Mercian Geol.* **2**, 267-278.

Smith, A. G. (1982). Late Cenozoic uplift of stable continents in a reference frame fixed to South America. *Nature* **296**, 400-404.

Smith, D. E., Kolenkiewicz, R., Dunn, P. J. and Torrence, M. H. (1979). The measurement of fault motion by satellite laser ranging. *Tectonophysics* **52**, 59-67.

Smith, P. J. (1974). Changing views of mantle viscosity. *Nature* **252**, 99-100.

Steers, J. A. (1945). *The Unstable Earth* (1st edn 1932). Methuen, London.

Stein, R. S. and King, G. C. P. (1984). Seismic potential revealed by surface folding: 1983 Coalinga, California, earthquake. *Science* **224**, 869-871.

Steinbrugge, K. V., Zacher, E. G., Tocher, D., Whitten, C. A. and Claire, C. N. (1960). Creep on the San Andreas Fault. *Bull. Seism. Soc. Am.* **50**, 389-415.

Stephansson, O. and Carlsson, H. (1980). Seismotectonics in Fennoscandia. In: *Earth Rheology, Isostasy and Eustasy* (Mörner, N.-A., Ed.), pp. 327-337. Wiley, Chichester.

Stevens, G. (1974). *Rugged Landscape.* Reed, Wellington.

Tapponnier, P., Mercier, J. L., Armijo, R., Tonglin, H. and Ji, Z. (1981). Field evidence for active normal faulting in Tibet. *Nature* **294**, 410-414.

Tchalenko, J. S. (1975). Strain and deformation rates at the Arabia/Iran plate boundary. *J. Geol. Soc. Lond.* **131**, 585-586.

Thatcher, W. (1976). Episodic strain accumulation in southern California. *Science* **194**, 691-695.

Thatcher, W. (1984). The earthquake deformation cycle, recurrence, and the time-predictable model. *J. Geophys. Res.* **89**, 5674-5680.

Thatcher, W. and Matsuda, T. (1981). Quaternary and geodetically measured crustal movements in the Tokai District, Central Honshu Japan. *J. Geophys. Res.* **86**, 9237-9247.

Thommeret, Y., Thommeret, J., Laborel, J., Montaggioni, L. F. and Pirazzoli, P. A. (1981). Late Holocene shoreline changes and seismotectonic displacements in western Crete (Greece). *Z. Geomorph.* (Suppl.) **40**, 127-149.

Thommeret, Y., King, G. C. P. and Vita-Finzi, C. (1983). Chronology and development of the 1980 earthquakes at El Asnam (Algeria): a postscript. *Earth Planet, Sci. Lett.* **63**, 137-138.

Thurber, D. L. (1972). Problems of dating non-woody material from continental environments. In: *Calibration of Hominoid Evolution* (Bishop, W. W. and Miller, J. A., Eds), pp. 1-17. Scottish Academic Press, Edinburgh.

Torge, W. (1980). *Geodesy.* de Gruyter, Berlin.

Torge, W. (1981). Gravity and height variations connected with the current rifting episode in northern Iceland. *Tectonophysics* **71**, 227-240.

Trifonov, V. G. (1978). Late Quaternary tectonic movements of western and central Asia. *Bull. Geol. Soc. Am.* **89**, 1059-1072.

Tryggvason, E. (1974). Vertical crustal movement in Iceland. In: *Geodynamics of Iceland and the North Atlantic Area* (Kristjansson, L., Ed.), pp. 263-275. Reidel, Dordrecht.

Tsubokawa, J., Ogawa, Y. and Hayashi, T. (1964). Crustal movements before and after the Niigata earthquake. *J. Geod. Soc. Japan* **10**, 165-6.

Turcotte, D. L. and Schubert, G. (1982). *Geodynamics.* Wiley, New York.

Umbgrove, J. H. F. (1942). *The Pulse of the Earth.* Nijhoff, The Hague.

UNESCO (1972). *The Surveillance and Prediction of Volcanic Action.* UNESCO, Paris.

Vail, P. R. and Hardenbol, J. (1979). Sea-level changes during the Tertiary. *Oceanus* **22**, 71-79.

Vail, P. R., Mitchum, R. M., Jr and Thompson, S., III (1977). Seismic stratigraphy and global changes of sea level, part 4. In: *Seismic Stratigraphy.* (Payton, C. E., Ed.), *Mem. Am. Assoc. Petrol. Geol.* **26**, 83-97.

Valentin, H. (1953). Present vertical movements of the British Isles. *Geog. J.* **119**, 299-305.

Vali, V. and Bostrom, R. C. (1968). The thousand meter interferometer. *Rev. Sci. Instrum.* **39**, 1304-1306.

van Bemmelen, R. W. (1954). *Mountain Building.* Nijhoff, The Hague.

van Bemmelen, R. W. (1972). *Geodynamic Models.* Elsevier, Amsterdam.

Vanicek, P. and Nagy, D. (1981). On the compilation of the map of contemporary vertical crustal movement in Canada. *Tectonophysics* **71**, 75-86.

Vita-Finzi, C. (1969). *The Mediterranean Valleys.* Cambridge University Press, Cambridge.

Vita-Finzi, C. (1973). *Recent Earth History.* Macmillan, Basingstoke.

Vita-Finzi, C. (1978). *Archaeological Sites in their Setting.* Thames and Hudson, London.

Vita-Finzi, C. (1979*a*). Contributions to the Quaternary geology of south Iran. *Rep. Geol. Surv. Iran.* **47**, 1-52.

Vita-Finzi, C. (1979*b*). Rates of Holocene folding in the coastal Zagros near Bandar Abbas, Iran. *Nature* **278**, 632-634.

Vita-Finzi, C. (1981). Late Quaternary deformation on the Makran coast of Iran. *Z. Geomorph.* (Suppl.) **40**, 213-226.

Vita-Finzi, C. (1982). Recent coastal deformation near the Strait of Hormuz. *Proc. Roy. Soc. Lond.* **382** A, 441-457.

Vita-Finzi, C. (1983). First-order ^{14}C dating of Holocene molluscs. *Earth Planet. Sci. Lett.* **65**, 389-392.

Vita-Finzi, C. and Ghorashi, M. (1978). A Recent faulting episode in the Iranian Makran. *Tectonophysics* **44**, 121-125.

Vita-Finzi, C. and King, G. C. P. (1985). The seismicity, geomorphology and structural evolution of the Corinth area of Greece. *Phil. Trans. Roy. Soc. Lond.* **314** A, 379-407.

Wadge, G., Horsfall, J. A. C. and Brander, J. L. (1975). Tilt and strain monitoring of the 1974 eruption of Mount Etna. *Nature* **254**, 21-23.

Walcott, R. I. (1973). Structure of the Earth from glacio-isostatic rebound. *Annu. Rev. Earth Planet. Sci.* **1**, 15-37.

Wallace, R. E. (1970). Earthquake recurrence intervals on the San Andreas Fault. *Bull. Geol. Soc. Am.* **81**, 2875-2890.

Wallace, R. E. (1977). Profiles and ages of young fault scarps, north-central Nevada. *Bull. Geol. Soc. Am.* **88**, 1267-1281.

Wallace, R. E. and Schulz, S. S. (1983). Aerial views in color of the San Andreas Fault. *US Geol. Surv. Open-File Rep.* 83-98.

Wang, C., Shi, Y. and Zhou, W. (1982). Dynamic uplift of the Himalaya. *Nature* **298**, 553-556.

Wegener, A. (1929). *The Origin of Continents and Oceans* (English translation of 4th revised edition, published in 1966). Dover, New York.

Weissel, J. K., Anderson, R. N. and Geller, C. A. (1980). Deformation of the Indo-Australian plate. *Nature* **287**, 284-290.

Wellman, H. W. (1971). Holocene tilting and uplift on the Glenburn coast, Wairarapa, New Zealand. *Bull. Roy. Soc. NZ* **9**, 221-223.

Wellman, P. (1981). Crustal movement determined from repeat surveying — results from southeastern and southwestern Australia. *J. Geol. Soc. Aus.* **28**, 311-321.

Wesson, R. L. and Wallace, R. E. (1985). Predicting the next great earthquake in California. *Scient. Am.* **252**, 23-31.

Westaway, R. and Jackson, J. A. (1984). Surface faulting in the southern Italian Campania-Basilicata earthquake of 23 November 1980. *Nature* **312**, 436-439.

Wetzel, R. and Morton, D. M. (1959). Contribution à la géologie de la Transjordanie. *Notes Mém. Moyen-Orient* **7**, 95-191.

White, R. S. and Klitgord, K. (1976). Sediment deformation and plate tectonics in the Gulf of Oman. *Earth Planet. Sci. Lett.* **32**, 199-209.

White, R. S. and Louden, K. E. (1982). The Makran continental margin: structure of a thickly sedimented convergent plate boundary. In: *Studies in Continental Margin Geology.* (Watkins, J. S. and Drake, C. L., Eds). *Mem. Am. Assoc. Petrol. Geol.* **34**, 499-518.

Whitten, C. A. and Claire, C. N. (1960). Analysis of geodetic measurements along the San Andreas Fault. *Bull. Seism. Soc. Am.* **50**, 404-415.

Williams, P. W. (1982). Speleothem dates, Quaternary terraces and uplift rates in New Zealand. *Nature* **298**, 257-259.

Wilson, L. G. (Ed.) (1970). *Sir Charles Lyell's Scientific Journals on the Species Question.* Yale University Press, New Haven and London.

Wu, P. and Peltier, W. R. (1983). Glacial isostatic adjustment and the free air gravity anomaly as a constraint upon deep mantle viscosity. *Geophys. J. Roy. Astron. Soc.* **74**, 377-449.

Wu, P. and Peltier, W. R. (1984). Pleistocene deglaciation and the Earth's rotation. *Geophys. J. Roy. Astron. Soc.* **76**, 753-791.

Wyllie, P. J. (1971). *The Dynamic Earth.* Wiley, New York.

Wyss, M. (Ed.) (1975). *Earthquake Prediction and Rock Mechanics.* Birkhäuser, Basel.

Wyss, M. (1977). Interpretation of the Southern California uplift in terms of the dilatancy hypothesis. *Nature* **266**, 805-808.

Wyss, M. and Baer, M. (1981). Seismic quiescence in the Western Hellenic Arc may foreshadow large earthquakes. *Nature* **289**, 785-787.

Yielding, G., Jackson, J. A., King, G. C. P., Sinvhal, H., Vita-Finzi, C. and Wood, R. M. (1981). Relations between surface deformation, fault geometry, seismicity, and rupture characteristics during the El Asnam (Algeria) earthquake of 10 October 1980. *Earth Planet. Sci. Lett.* **56**, 287-304.

Yorke, R. A. and Little, J. H. (1975). Offshore survey at Carthage, Tunisia, 1973. *Int. J. Naut. Arch. Underwater Explor.* **4**, 85-101.

Zak, I. and Freund, R. (1966). Recent strike slip movements along the Dead Sea Rift. *Israel J. Earth Sci.* **15**, 33-37.

Zak, I. and Freund, R. (1981). Asymmetry and basin migration in the Dead Sea Rift. *Tectonophysics* **80**, 27-38.

Zeitler, P. K., Johnson, N. M., Naeser, C. W. and Tahirkheli, R. A. K. (1982). Fission-track evidence for Quaternary uplift of the Nanga Parbat region, Pakistan. *Nature* **298**, 255-257.

Zeuner, F. E. (1958). *Dating the Past* (4th Edn.). Methuen, London.

Zoback, M. D. (1983). A new data source for *in situ* stress field orientations. *Nature* **306**, 18.

Zuchiewicz, W. (1984). Neotectonic movements in the Carpathians. *Tectonophysics* **104**, 195-204.

INDEX